A Life for Water
A Memoir

Luna B. Leopold.

A Life for Water
A Memoir

Luna Bergere Leopold
1915–2006

American Philosophical Society Press
Philadelphia

> Transactions of the
> American Philosophical Society
> Held at Philadelphia
> For Promoting Useful Knowledge
> Volume 108, Part 4

Copyright © 2020 by the American Philosophical Society for its Transactions series.

All rights reserved.

ISBN: 978-1-60618-084-6
US ISSN: 0065-9746

Library of Congress Cataloging-in-Publication Data

Names: Leopold, Luna B. (Luna Bergere), 1915-2006, author. | Leopold, Madelyn, editor. | Vita-Finzi, Penelope, editor. | Vita-Finzi, Claudio, editor.
Title: A life for water : a memoir / by Luna Bergere Leopold ; [edited by] Madelyn Leopold, Penelope Vita-Finzi, and Claudio Vita-Finzi. Description: Philadelphia, [Pennsylvania] : American Philosophical Society Press, 2019. | Includes bibliographical references and index. | Summary: "This is a memoir by Luna Leopold, chief hydraulic engineer and later chief hydrologist for the U.S. Geological Survey"— Provided by publisher.
Identifiers: LCCN 2019031902 (print) | LCCN 2019031903 (ebook) | ISBN 9781606180846 (paperback) | ISBN 9781606180891 (kindle edition)
Subjects: LCSH: Leopold, Luna B. (Luna Bergere), 1915-2006. | Hydrology—Research—Biography. | Hydrologists—Biolography. | Water conservation—Research—Biography.
Classification: LCC GB659.72.L46 L46 2019 (print) | LCC GB659.72.L46 (ebook) | DDC 551.48092 [B]—dc23
LC record available at https://lccn.loc.gov/2019031902
LC ebook record available at https://lccn.loc.gov/2019031903

Contents

The Career of Luna Leopold   ix
Foreword   xiii
Preface   xv

1. **Precious Water**   1
    Benefits and Costs   3

2. **Learning to Observe**   7
    The Skill of Estimating   8
    Need for Trust   10
    Opportunity   12
    Interruption by War   14

3. **Research on a Practical Problem**   19
    Meteorology as a Profession   22
    Observing Natural History   24
    Kauai   25
    Local Fishing   32
    Agriculture   32
    Molokai Landscapes   33
    Crabs, Fish, and Research   37
    Views from the Coastal Cliffs   40
    A Chance Observation That Changed the Direction
        of My Research   44
    Ranching on the Big Island   44
    Water Supplies   47
    The Rainfall of East Maui and Research
        in Climatology   49

4. Travels Toward Science   57
Terraces of the Powder and the Bighorn Range   61
The New Fork and the Green River   70
A Plan for Research   79
Hydraulic Geometry   81
Flood-Control Problems   83
India: Water Development Viewed from the Inside   83
An Engineering Office   86
Poona Experiment Station   88
Looking for Data   91
Problems of Development   97
Plans Put on Hold   102
Nails in the Ground and Tragedy   104
Pursuit of Our Aims   105

5. Science as a Craft   109
Framing the Questions   110
A Plan for the Work   116
The Gates of Lodore   119
Cataract Canyon of the Colorado   125
Russia   132
Ancient Sources and Developments   146

6. Some Water Studies   155
The Concept of *Base Level*   157
The Loop of the San Juan   159
Mesa Verde   161
The Negev, Israel   162
Effect on Geologic Thought   163
River Meanders   164
The Grand Canyon   170
The Arctic Plain and Kotzebue   180
The Mackenzie and Inuvik   188
Two Wild Rivers   193
Comparison of Two Wild Rivers   201

CONTENTS

7. Environmental Impact   205
    The Florida Jetport and the First Environmental
        Impact Statement   206
    The Oil Pipeline in Alaska   211

8. Environmental Ethics   219
    Values in Conservation   220
    A Reverence for Rivers   226
    Let Rivers Teach Us   234
    Ethos and the Earth's Resources   237
    Government and the Loss of Equity   240
    The Hydrologic Continuum   243

Bibliography   247

Index   251

The Career of Luna Leopold

Born in Albuquerque, NM, 8 September 1915

Education

1936 BS Civil engineering (University of Wisconsin, Madison)
1944 MS Physics—meteorology (UCLA)
1950 PhD Geology (thesis on The Erosion Problem of the Southwest; Harvard University)

Career

1937–1940 Soil Conservation Service
1940–1946 US Army Weather Service
1946–1950 Pineapple Research Institute, Hawai'i
1950–1970 Water Resources Division, United States Geological Survey (USGS)
1972–1986 Departments of Geology & Geophysics and Landscape Architecture, University of California at Berkeley

Major Honors and Awards

1958 Kirk Bryan Award, Geological Society of America
1958 Distinguished Service Medal, U.S. Department of the Interior
1968 Member, National Academy of Sciences
1969 Fellow, American Academy of Arts and Science
1972 Rockefeller Public Service Award
1972 American Philosophical Society
1973 Warren Prize, National Academy of Sciences, United States
1982 Fellow, California Academy of Science
1983 Busk Medal, Royal Geographical Society London
1991 National Medal of Science

1993 Robert E. Horton Medal, American Geophysical Union
1994 Penrose Medal, Geological Society of America
2006 (jointly with M. G. Wolman) Benjamin Franklin Medal

Key Publications

1937 Relation of Watershed Conditions to Flood Discharge: A Theoretical Analysis. *Bulletin of the U.S. Department of Agriculture*, no. 57, 22 pp.
1944 Characteristics of Heavy Rainfall in New Mexico and Arizona. *Transactions of the American Society of Civil Engineers, 109*, 837–66.
1951 Pleistocene Climate in New Mexico. *American Journal of Science, 249*, 152–68.
1951 Vegetation of Southwestern Watersheds in the Nineteenth Century. *Geographical Review, 41*, 295–316.
1953 (with T. Maddock, Jr.) The Hydraulic Geometry of Stream Channels and Some Physiographic Implications. *U.S. Geological Survey Professional Paper 252*, 56 pp.
1954 (with T. Maddock, Jr.) *The Flood Control Controversy.* New York: The Ronald Press, 278 pp.
1954 (with J. P. Miller) A Postglacial Chronology for Some Alluvial Valleys in Wyoming. *U.S. Geological Survey Water-Supply Paper 1261*, 100 pp.
1956 (with J. P. Miller) Ephemeral Streams: Hydraulic Factors and Their Relation to the Drainage Net. *U.S. Geological Survey Professional Paper 282-A*, 38 pp.
1956 Land Use and Sediment Yield. In *Man's Role in Changing the Face of the Earth*, edited by William L. Thomas, Jr., 639–47. Chicago: University of Chicago Press.
1957 (with M. G. Wolman) River Channel Patterns: Braided, Meandering, and Straight. *U.S. Geological Survey Professional Paper 282*, 52 pp.
1960 (with R. A. Bagnold, M. G. Wolman, and L. M. Brush) Flow Resistance in Sinuous and Irregular Channels. *U.S. Geological Survey Professional Paper 282 D*, 111–34.
1962 (with W. B. Langbein) The Concept of Entropy in Landscape Evolution. *U.S. Geological Survey Professional Paper 500 A*, 20 pp.

1962 (with R. L. Nace) Government Responsibility for Land and Water: Guardian or Developer? In *Land and Water Use,* 349–57. Philadelphia: American Association for the Advancement of Science.

1962 The Vigil Network. *International Association of Scientific Hydrology,* 7, 5–9.

1964 (with M. G. Wolman and J. P. Miller) *Fluvial Processes in Geomorphology.* San Francisco: W. H. Freeman, 522 pp.

1966 (with W. B. Langbein) River Meanders. *Scientific American,* 214, 60–70.

1967 Rainfall Frequency—An Aspect of Climatic Variation. *Transactions of the American Geophysical Union,* 32, 307–357.

1978 (with T. Dunne) *Water in Environmental Planning.* San Francisco: W. H. Freeman, 818 pp.

1979 (with W. B. Bull) Base Level, Aggradation and Grade. *Proceedings of the American Philosophical Society,* 123, 163–202.

1980 Bathymetry and Temperature of Some Glacial Lakes in Wyoming. *Proceedings of the National Academy of Sciences,* 77, 1754–1758.

1988 (with D. C. Rosgen) Natural Morphology, Key to Channel Stability. In *Proceedings, International Mountain Watershed Symposium,* edited by I. G. Popoff et al., 42–50. South Lake Tahoe, CA: Tahoe Resource Conservation District.

1990 Ethos, Equity and the Water Resources. *Environment,* 32, 16-41.

1994 *A View of the River.* Cambridge, MA: Harvard University Press, 290 pp.

1997 (with W. W. Emmett) Bedload and River Hydraulics—Inferences from the East Fork River, Wyoming. *U.S. Geological Survey Professional Paper 1583,* 52 pp.

1999 (with I. Lucchitta) Floods and Sandbars in the Grand Canyon. *GSA Today,* 9, no. 4, 1–8.

2000 Temperature Profiles and Bathymetry of Some High Mountain Lakes. *Proceedings of the National Academy of Sciences,* 97, 6267– 6270.

2004 A Sliver Off the Corpus of Science. *Annual Review of Earth and Planetary Sciences,* 32, 1–12.

Foreword

Over the course of his long scientific career, Luna Leopold wrote occasional essays of a personal nature. This volume presents a selection of those essays that we think usefully complements his articles and books. In particular, they show how he became increasingly concerned with environmental degradation and with unethical use of the land. There is an obvious link to the work of his father, Aldo Leopold, often considered the founder of the conservation movement, but the differences are significant. Luna viewed environmental issues from the standpoint of the engineer rather than the ecologist, and his work as a hydrologist, perhaps the foremost student of rivers of the 20th century, ensured that his proposed solutions were practical as well as ethically framed.

We have edited these essays very slightly, mainly by correcting obvious typographical or topographical errors, eliminating repetition, completing personal names, and clarifying obscure references. Our intention is to respect Luna's distinctive voice, or perhaps voices, as what follows ranges from the dry detachment of the bureaucrat to the slangy bark of the hunter and pilot. Luna's formal publications, journals, and correspondence are archived at the Library of the American Philosophical Society in Philadelphia. We thank Pamela Lankas of IGS and Mary McDonald, Director of Publications at the American Philosophical Society, for their enthusiasm and patience in the preparation of this volume.

The Editors

Preface

One might suppose that a person whose main interest and professional writing concerned a field like water would have spent most of his or her life dealing with topics that involved water. Such a view is quite wrong. In fact, most of one's life is dominated by much more mundane things, such as holding a job, maintaining a family, and meeting social and professional obligations. Indeed, a scientific career comprises mostly learning, reading, observing, and recording.

Viewed in this light, it is not surprising that a book about a life in science deals mostly with observing and learning. This is just a personal memoir, but it is still of some interest to see how detailed observations were influential in contributing to a larger view of the author's principal concern, in this instance water on the earth.

Water is so embedded in all aspects of the earth—in physical and chemical processes, the biota, earth history, and much more—that a lifetime of observation and study merely touches the surface of what there is to know.

In a decade central to my life, I was immersed in the administration of water studies in my position as Chief Hydraulic Engineer and later Chief Hydrologist of the United States Geological Survey. It was an important period in the study of water for I was in a position to initiate and oversee a greatly expanded program of research and data collection, enhanced by my determination to join the previously disparate subjects of geomorphology and hydrology. Some of the activity described in this book took place during that critical period of time, including the launch of the Professional Paper Series of the Geological Survey, which was to become one of the most respected and prominent sources of expanded knowledge of water on the earth.

The last portion of the book departs from the form of a personal memoir and is a series of essays on water and modern politics, which is influencing not only the water resource but all the resources of the earth, to our great disadvantage.

The depth of experience and pleasure that I have had in my career has been greatly enhanced by colleagues and friends, many of whom are mentioned in this volume.

Among those are individuals whose lives as well as mine were devoted to understanding aspects of water science: Smuss Allen, Howard Chapman, Joshua Collins, Laurel Collins, David Dawdy, Thomas Dunne, William W. Emmett, Scott McBain, John P. Miller, J. David Rogers, David Rosgen, Herbert E. Skibitzke, and M. G. Wolman. For ideas, clarification, and assistance, I am grateful to Paul Bierman, C. L. Rawlins, William Trush, and an unknown reviewer. For his important help in manuscript preparation I thank Raymond Ulibarri.

1
Precious Water

As I start writing what I hope might eventually be published, I look out at the little stream that crosses the yard of my small property in Wyoming. My stream is very low, for it has been progressively depleted by my neighbors upstream, who need the water for irrigating their hay meadows. I need the water for my few trout and just for looking at it. Beyond are the peaks of the Wind River Range, still partly snow covered in July, and I like to see these framed in the pastoral landscape.

I feel lucky to have any water at all in my river and what I get is just return flow from the irrigated pastures upstream, for there is no intention by others to leave water in the channel to satisfy my esoteric desires. But there is water and there are many people in the world who have even less. Thus, I can see in this one spot a panoply of natural resources of the earth and some of the results of the needs, desires, ethos, and actions of humans utilizing the bounty of nature.

When we in America turn on the tap to draw some water, we don't think at all about the fact that this is a near miracle. For most of the world it would indeed, indeed, be a big thing. We quite forget, even if we knew it once, millions of people do not have water to serve their basic needs, access to potable water and sanitation. Experience has shown that the minimum amount of water needed each day for each person is 13 gallons, or 50 liters, to be used for cooking, washing, bathing, health needs, waste disposal and sanitation. In the United States we are presently using 100 gallons per person per day, or 380 liters. There are 1.3 million people in the world who do not have that minimum of 50 liters.

It has long been assumed that the way out of this dilemma was to develop more water by the usual engineering means: building dams, tapping more groundwater and diverting rivers that flow from areas that have good supplies to those areas that are short of water.

We are not going to be able to fill this gap by the development of new supplies. The only viable option is to make better and more efficient use of the water supplies already available to us: to meet the needs of the people by supplying the amount, quality, and source appropriate to the actual end-use. It means equitable access to water among different users, proper application and use of economies, incentives for efficient use, delivery reliability, and public participation in decision-making.

The previous reliance on structures to store, divert water, or produce hydroelectricity, always had associated damages that until recently have been unacknowledged, undervalued, and ignored. The change of the amounts and timing of flow caused by dams or diversions always alter the ecology of the river downstream. To people who have depended on fish or other river-related assets, these changes have been catastrophic. The construction of a dam involves creation of a reservoir that often displaces people who have lived there, and though governments promise to provide new living space, it always is lower in value, productivity, or convenience. The breakup of community, customs, and modes of living cannot be cancelled. The drowning of religious and historic sites is a severe blow to the society. These are real losses but are usually impossible to quantify and cannot even been expressed in monetary terms.

Until the last decade, these effects have not been listed, explained, or publicized and even now are summarized by only a few publications like the *World Rivers Review* or reports of the Commission on High Dams. Unfortunately, these destructive practices on nearly all continents are financed with programs for public assistance from the World Bank, the Export–Import Bank and others, most of which ignore the kinds of widespread damage and misery that they are promoting.

Benefits and Costs

For several decades it has been supposed that a facility, a structure, a dam or other entity can be judged by a computation of the cost–benefit ratio. If the computed benefits exceed the estimated cost, it is reasoned that the structure or procedure is justified. This reasonable-sounding rule might be useful if both sides of the equation could be properly evaluated. Experience has shown in a large number of cases even the direct cost of construction or action is minimized or underestimated while the benefits are consciously exaggerated. Further, the indirect costs that accrue from associated forms of damage are simply ignored, and especially in aesthetic or environmental conditions that the public values but for which monetary statements are invalid, inappropriate, or meaningless.

To give some simple examples, a national park exists because humans desire the opportunity to experience natural beauty, solitude, wilderness, presence of wildlife, and opportunity for sports. There are monetary advantages to businesses that sell food, equipment, services and goods that depend on the presence of the park. But the sum of monetary transactions of all these does not reflect the value of the park. That value is not expressed in monetary terms. Further, the value of a park is not measured by the number of visitors it has. As use increases, the opportunities for solitude and other aesthetic or intangible values decrease.

Meeting the growing demand for water calls for a new mindset that goes beyond ecology and that recognizes not only the natural sciences but also technology and sociology. To achieve a general respect for landscape there must be a realistic strategy for satisfying human aspirations. The approach requires a recognition of the many and diverse ways water affects human affairs and human needs. It requires knowledge, data, and analytical tools, including a gradually expanding understanding of how water is influential in forming and maintaining a landscape. Such understanding comes from scientific study of ecology, geomorphology, geography, geology, plant physiology, taxonomy, and climatology as well as a variety of the social sciences.

It is in these sciences and their relation to water to which I have devoted my career. In addition to the scientific aspects of these subjects, I am particularly interested in the intangible values, obscure processes in the water environment, and the origin and causes of water-related phenomena. My special concern is with rivers. I had long experience in developing and improving the capacity to measure different aspects of water, quantity, quality, water science, and water use.

I think it is true that some of the best research comes out of experience with the practical problems facing individuals and organizations. Or perhaps it is better to say that immersion in large practical problems gives stimulation and ideas to a potential researcher. This book is an attempt to show how that combination worked for me and, indeed, when I was at the helm of a large organization in the field of water, it was the most productive feature of my research career. A description of how some ideas originated and the sequence of their development may give the reader an insight into how research and experience combined at least in one individual's life.

To explain how ideas develop and how they fit into the ordinary aspects of one's life is, perforce, autobiographical. I hope that a description of personal experience will serve to illustrate how research and the skills that go into it can be cultivated. Some insight also comes from observing other scientists or reading about their discoveries.

2
Learning to Observe

The Skill of Estimating

The summers during five years of college studying engineering and biology were spent at agricultural experiment stations. The first summer in 1932 was at the Coon Valley Station at La Crosse, Wisconsin, the following year at the Sierra Ancha Station of the US Forest Service near Globe, Arizona, and then the Navajo Experiment Station of the Soil Conservation Service near Nakaibito, New Mexico. In the first two I was a kind of intern without salary but at Nakaibito I was a young engineer. For a time I was a rodman for Mr. Herbert Yeo, formerly State Engineer of New Mexico, an interesting character with a wealth of experience. We were making a map of a reservoir site by plane table. Quite often, Mr. Yeo would say to me, "How far is it to that tree?"

"I don't know sir."

"Well, estimate it," which I did very roughly.

"No," he said, "it is 221 feet." He always made his estimate sound as if it were highly accurate, all to make a point of course, but indeed he was very proficient. One day my guess was better than his and he said, "All right, I'll hold the rod for you and you run the instrument."

On another occasion we stopped on a bridge overlooking a dry streambed in a deeply incised channel. He asked me to estimate the discharge in cubic feet per second in the last flood in the channel. Of course, I could not do it, but he explained in detail the procedure step by step.

I was learning a highly useful skill. Later I was told to make a plane table map of every plant in a 100-meter fenced quadrat, that is a square 100 meters on a side. The purpose is to see how the plants grow when they are released from grazing. These fenced plots have been very useful and sometimes controversial because the scientist does not necessarily know how to interpret the result in terms of management. My assignment was to identify each plant, note its size or condition and locate it on the map I was making. That was made much easier if the distance to each plant was estimated before it was measured by stadia and that is where the experience with Mr. Yeo was so useful. As it turned out, I was to come back to this place years later, in 1946, to look for the quadrat.

Another assignment that summer was to supervise and assist a team, including several Navajos, building check dams in the arroyos on the experiment station. These small dams were made of rock without mortar, sometimes using wire or fencing materials to stabilize the structure. I was not convinced of either stability or the long-term effect of these dams and voiced this to the engineer in charge of that operation. The conversation went something like this:

"You see," he said, "given time the deposition behind the dam will extend itself upstream and finally fill the whole gully. By that time vegetation will have grown up in the vicinity of the dam and when, finally, the dam fails, the vegetation will have stabilized the sediment deposit."

"But sir," I replied, "there is no indication of the sediment deposit extending beyond a short distance upstream, even in some relatively old structures."

"Time has been insufficient. Give them some more years."

This is essentially the problem of base level, one on which I have spent much time over the years and is the subject of one of the scientific problems discussed in a later chapter.

It was during these years of late high school and college that I spent much time hunting, especially for upland game, snipe, doves, partridge, quail, and on occasion sharp-tailed grouse. I did not care much for duck hunting. For generations the men in the Leopold family were hunters and fishermen. It is a perfect way to teach restraint, ethical conduct, and good manners. But in that period our father, Aldo, became less interested in shotguns and more concerned with the long bow. He started making his own tackle for he was a skilled craftsman in wood, and the two boys, Starker and I, followed suit. In time, we also became good artisans. So, all three of us made bows and arrows, Aldo making the equipment for mother as well for himself. My mother, Estella, was the most talented of all and was a better archer than anyone. For some years she repeatedly won the women's championship in the state. Our Sunday routine was to go roving, meaning we would walk through the countryside, picking out things to shoot at like a clump of grass, a colored leaf, an anthill or a stump. It was good practice for the few times we went deer hunting, none of us ever got a deer but it was a great sport.

Need for Trust

After five years I graduated from engineering school in 1936, unusually well prepared for work in aspects of conservation in the resource field. My father, Aldo Leopold, a now famous ecologist, was not pleased that I was enrolled in civil engineering. He thought my way of thinking was better suited to work in agriculture. I explained to him that engineers were entering the field of land management and I felt that one can talk to engineers better if he were one of them. Dad asked questions but never told us what to do.

At that time it was easy to obtain a job quickly if I would accept a sub-professional position. Within a few months a scheduled examination was held so with a high score I was advanced to the professional rank of civil engineer. That first year was a learning experience, for I was able to work closely with an interdisciplinary team of men in range management, forestry, soils and geology, all older than I. Looking back at it, I was a real greenhorn, but I worked hard and learned. Our office was well ahead of others in that the officers realized the need for several or many disciplines working together to make progress in the field of land management.

The top man of the section where I was assigned was more administrator than technical man, but one who wanted to absorb as much professional knowhow as he could.

During this time, I became friendly with the geologist in our office, Dr. Parry Reiche, who treated me with respect and was very helpful to me in my work, even though I was not assigned to him. Through my association with Dr. Reiche, I also got to know his secretary. One day she said, "Why is it that you showed no interest in the several offers of a higher position?"

"Job offers?" I said, "I never saw a job offer."

"They were sent to your top boss, and it is clear he never showed them to you."

Here was a classic example of failure in all the usual attributes that make the relation of supervisor to subordinate viable. It was a breach in the code of trust, honesty, and ethical behavior. I was just beginning to appreciate the precepts and the importance of relationships.

I suggested to the head officer, the Regional Director, that we should know more about ideas of the well-known and outspoken geologist, Kirk Bryan, who had written extensively that the engineers in the land agencies were on the wrong track because they did not understand that climatic change would account for many of the erosion problems the land was experiencing. My supervisors considered it a bad idea and one coming from a puerile youth. Shortly I resigned and sought admittance to Harvard to study under Bryan. It was a great learning experience and quite different from my previous courses in engineering I could not continue because I had no money.

After that year of graduate study, 1937, I came back to the same office in Albuquerque, but had made in advance a request to work under an experienced engineer whose reputation had been growing as the expert in water issues, Thomas Maddock, Jr. We were assigned a small office furnished with a double desk and two chairs. I sat opposite my boss. He seemed continually to have his feet up on the desk reading, and what he read was a wide assortment of the newest works in hydraulics and in the developing science of hydrology, much of which was to be found in the *Transactions of the American Geophysical Union.*

Our joint assignment was to develop flood control plans for various ephemeral washes in the southwestern states. Thomas Maddock was a large man with a shock of unruly hair, a neat mustache, slow, western speech, very bright and serious eyes, and a great spirit of inventiveness and capacity for learning. He was some years older than I but came to be a lifelong friend.

A typical week started with an exchange:

"Say, Luna, here is a brand-new technique just published. It measures the movement of water under conditions rather similar to those we see in an arroyo. Let's try it on that tributary near Bernalillo."

"We have to have some rain data."

"Well, you go look for the needed data; go to it," he said. "Then we can begin the computation."

"I'll be back in a few days."

I found what I needed. He trusted that I could do the necessary work. He gave me confidence as well as guidance.

Those few years just before the war were very happy ones for me. I was learning new things; the technical work was interesting and fulfilling. Years later we looked back on this period and realized we were on the forefront of the newly developing science of hydrology. Tom devoured every new idea, concept, or procedure that appeared in print. We examined each with actual data, changed it, and often improved it. His reading habit extended to every aspect of the related sciences: hydraulics; hydrology; climatology; range management, especially grazing, agronomy, and meteorology. We dipped into them all. We traveled over much of New Mexico and Arizona, and maintained close contact with the Corps of Engineers personnel in California.

Opportunity

It was a great advantage to me to work under Tom Maddock. I carried out his wishes in the computations and plotted the results. This deserves more explanation. In 1938 the leaders in the field of hydrology were Robert E. Horton, W. W. Horner, and L. K. Sherman. As an engineer with long and varied experience, Horton was an expert in hydraulics and a keen observer of runoff events. He was at that time developing the theory of infiltration. Horner was studying runoff from urban areas, and Sherman was in the process of developing the concept of the unit hydrograph. We were reading everything we could find written by these important hydrologists, tried their procedures with our own data, and sought ways to apply them to definite local problems.

An opportunity arose for one of the junior staff to attend a field class in sediment deposition in small reservoirs. Tom immediately put my name forward and I was chosen. The work started in Oxford, Mississippi, where the senior scientists worked, especially Stafford Happ, whom I later supported to resurvey the work being done originally out of Oxford. I remember writing a poem called "A Cambridge Man in Oxford," a copy of which has long been lost.

The work was under the direction of Carl Brown, a man well known in the sediment field, and later a friend and associate. We were given field experience in sediment measurement in small lakes and ponds, and we traveled to various locations in the southern states. It was a great experience for which my thanks extend to my superior, Tom Maddock. He always took care of me and gave me every break.

A later and equally important assignment came up when the Washington office wanted a few engineers to gather to compute for the technical leaders, including Horton and Horner. Just prior to that assignment I had been in the field making measurements of infiltration, comparing various soils in the Pecos watershed. We used what was called a *Type A infiltrometer*, consisting of a small plot, about 1 x 2 meters, under a tent where the soil was sprinkled with water. The changing rate of runoff allowed us to compute the rate of water infiltrating into the soil.

The great engineer Robert Horton, as I mentioned, was developing his theory of infiltration so the data we were collecting in the Pecos was of great interest to him. I will never forget his imposing presence: tall, a shock of unruly white hair, an impressive voice, and a gentlemanly way of dealing with those beginners like me. That trip to Washington was notable in many ways for I met many interesting professionals with whom I had dealings over the years.

Back in Albuquerque in 1940, I continued my work on rainfall and runoff in many interesting and unusual events. I determined to reconstruct unmeasured storm events by looking at chance indicators. For example, in chasing a storm event it was my procedure to look for any container that might have held water from the rainstorm. A variety of odd cans, open jars, old hubcaps, or any other catchment would be used. From such scattered evidence and ground inspection I constructed an approximate map of the rainfall and the field study of the stream channels gave some estimate of runoff.

On one of these trips chasing storms, I passed on a lonely road a large black car on the side of the road. Noting that the driver was a woman, I pulled to a stop, backed up, and asked if I could help. The middle-aged woman and her young-adult daughter were glad to have some help, for the car had just plain stopped. I lifted the hood and to my amazement, I was looking at the 16-cylinder engine of a large Cadillac. Now I am no mechanic, but as an engineer I made some preliminary guesses, principal among them was that the engine was not getting any gasoline.

Not feeling very confident, I took out my small tool kit and took apart the gasoline tube leading to the carburetor. Freeing the tube I blew through it to the gas tank and then attempted to clean any filter I could find in the daunting carburetor. Then I reattached the gas line, got in the driver's seat, turned the key, and the damn motor purred. I was just as amazed as the two women were. They explained that they were on their way to central Mexico, but the car had died in southern Arizona. The lady observed the grease on my hands and said, "Here's something to clean your hands," and handed me a full bottle of Regis Scotch whiskey. I said I would use it for a swallow but not for cleaning my hand. Then she asked me if I would be their guest and drive them to Mexico. Overly righteous (a fault repeated more than once when I was young) I thanked her and said no. What a fool!

Interruption by War

It was a period of intense learning. Tom not only gave me a lot of leeway to do things I thought needed study but he recommended me for any learning experience that presented itself. He decided we should go to Los Angeles to talk with engineers we knew in the hydraulics sector. We flew in the earliest commercial airline in operation. The plane was a Ford Trimotor. Inside the plane there were metal benches on each side of the hull and we sat on these. I do not recall if there were other passengers. The matters we discussed with our friends all had to do with flood control. About that time the two of us started a plan for flood control on the Mesa above Albuquerque, the great expanse of sediment surface that extended from the Sandia Mountains to the city. There were no developments as later were built, only a few homes in the vicinity of the university campus. There was nothing but open country from the bluff overlooking the Rio Grande Valley all the way to the mountains.

This great open space is drained by a series of shallow ephemeral draws that originate at the mountain front and proceed as a coarse network of braided channels that finally cut rills in the scarp that bounded the river floodplain. We drew a design for flood control works that would surely be needed by the city at some time in the future, but I doubt if the plan was ever submitted to the city for time and events overcame it.

Pearl Harbor changed the focus for everyone. Most of us felt that there were more immediate tasks than those that occupied our time and efforts. I called or wrote to the contacts Tom and I had developed with the Corps of Engineers in Los Angeles. They assured me that if I came to them there were plenty of engineering jobs to be performed.

That was enough for me. I prepared to go to join the war effort. That seems now like a cliché, but at that time it was real.

We had been renting a small house in west Albuquerque on Tulane Place. The house was at the edge of the east side of the city with nothing between us and the Sandia mountains; a clear view. I had found on the floodplain of the Rio Grande in the fall a cottonwood tree about 4 inches in diameter that I marked as a male that would not shed the cotton in the spring that plugged the window screens to the dismay of housewives. In the spring I dug it out bare-rooted with all the roots cut off to mere stumps, and I pruned all the branches. But when put in the ground, it sent out new buds to my delight and amazement, considering how radically I had pruned it. But it grew.

About 1968 when I flew a light plane and passed over Albuquerque, I could see where I had lived because my tree was the largest to be seen on the east mesa. About 1985 I visited that house and looked at the tree I planted in 1941, and it was about 16 inches in diameter and just beautiful.

The office of the Corps of Engineers where I worked was on Figueroa St. near the center of the city. I was given a desk and told to oversee contracts with consulting engineers doing work for the War Department. Such contracts included sewage disposal plants, buildings, and utilities. I soon learned the special protocol of written documents in that arm of government. The secretaries were pretty and very good at shorthand so I got experience in dictation. Soon I was given more responsibility and left to do my work with minimum supervision. I obtained more experience at what it was like to work under a good boss, Clarence Shidel, an engineer I admired greatly.

I was then put to work designing a training center in the Mojave desert where troops were to learn desert warfare in tanks and armored vehicles. For this kind of an installation, the specifications called for a flagpole in front of the command center. The flagpole was to be some 50-feet high and made out of laminated wood in the form of a round pole tapering toward the top. I thought that during wartime such a pole was unnecessary and too expensive, so I called for a well-chosen telephone pole. This was the only time the officer in charge in my office confronted me with an order to change my decision. He said, "The commanding officer of the camp will not be satisfied with a telephone pole."

"Why not?" I asked.

"Well, the officer is General Patton." I changed my order.

Being a civilian in an office run by an officer, even if he is of low rank, is quite an experience. I wanted to be advanced to an officer but was told that there was no way in the army that a civilian could be so advanced. The officers who I knew had come at it through the National Guard.

My experience with the relation of supervisor to worker was expanding. So, I went progressively to the Navy and the Marine Corps to see if they needed an engineer in the service. Both said they would be glad to accept me as an officer to be put directly in a field assignment. After seeing the relative opportunities, it was clear that the Marines wanted engineers to build bridges for troops and the Navy would assign one to water-supply work. Because I had no experience in bridges, but training and fieldwork in water science, I chose the Navy. I took the physicals, had the shots, fulfilled other requirements, and was ready to be sworn in.

The swearing ceremony was scheduled for 8 a.m. on a Monday at the Federal building in Los Angeles. I had told the Corps of Engineers my decision and the officers accepted my departure. On the appointed Monday I got to the Federal building at 7:30, wishing to be prompt. The Federal building is a massive structure approached by a long flight of stairs, and I watched workers arriving for duty and climbing the long flight of steps. Particularly I inspected the Navy officers who, to a man, climbed the steps in proud uniforms and at the top, each turned smartly around and saluted. I kept watching this procedure, trying to figure out what they were saluting. I finally realized that they were saluting the flagpole, but there was no flag. Salute the pole without a flag? I turned on my heel, walked down the steps, and went along a main street till I came to a recruiting station. I told the sergeant I wanted to enlist. Within days, I was in boot camp as a private, Army of the United States.

My most vivid memory of those first weeks was the picture of long lines of nude men waiting for shots. The non-coms in the medical unit were pretty impersonal and didn't change expression when a man gasped, moaned, or just swore when the needle was dull and the shot just plain hurt. The physical routine of running, climbing, and passing hurdles was tough but nothing like what came later.

We collected our clothes from a warehouse and fitted heavy ankle-high shoes, and soon learned the initial routine of marching and all the commands. For some of us, more important was the opportunity to apply for officer candidacy. There were several but the only one important to me was the possibility of expanding my beginning knowledge of hydrology. Because I was trained in the engineering part of water science, I saw that to be trained in meteorology would be an excellent expansion for me. I made an application for this opportunity. It was known that the army was training people at Chicago, Massachusetts Institute of Technology, and at the University of California at Los Angeles. I seemed to have the necessary qualifications, a B.S. in civil engineering that included mechanics, physics, and math.

I was accepted and assigned to University of California Los Angeles Physics Department as a cadet. Most cadets, however, came from mathematics or physics and had better background than I for this work. Our class consisted of about 110 cadets, all housed in apartment buildings not far from the center of Westwood, three men to a room. Our instructors were mostly persons who had taken the same course previously, not military personnel. Most cadets were slightly younger than I and more recently out of school. In that situation I had a difficult time because the classwork was mostly mathematical and the physical training very strenuous. On the first of the many class examinations I received the grade of 62. I was sure I was going to flunk out as well as die from exertion. One night I was so agitated my heart seemed to be pounding in an unusual way. I had one of my roommates call an ambulance and I was whisked to an army hospital. The physician told me it was nerves, go back, and keep on trying.

As the year progressed, I began to feel more confident, especially because I was good at the synoptic map analyses and certainly learned a lot of physics that I had either never known or forgotten. The faculty, permanent and temporary, treated the cadets in a gentlemanly manner, serious but helpful, impersonal but friendly. It was greatly appreciated.

It was a difficult but satisfying year. When we marched out in close formation into the center of the Los Angeles Coliseum, there to receive our bars as commissioned officers, even the most sardonic among us were impressed. We got orders to be assigned to different units at home or abroad, and it was an anxious time waiting for the papers off the teletype. For a few officers, I among them, the orders were to the command at UCLA to teach the next class of cadets.

Teaching cadets was an experience in tempering hard-headed discipline with the sympathy that came from having gone through what these students were experiencing. I was acutely conscious of the stress because I had started out so poorly at the beginning of the term.

3
Research on a Practical Problem

When the year of teaching ended, the whole corps of lieutenant teachers was sent to San Antonio, Texas. Again, at the end of the term, we went through the same anxiety of waiting for orders because the time in Texas dragged on to three weeks. When I finally got my orders it was, to my surprise, to return to UCLA to carry out research in weather forecasting. I was made commanding officer of a small unit consisting of two officers and five enlisted men.

The weather in Los Angeles is typically Mediterranean, with wet winters and long dry summers, but with the added complications of a strong sea breeze, low stratus clouds that burn off in late morning giving way to sunny afternoons. The depth and strength of these low clouds changed day to day and forecasting the onset and dissipation of the clouds and their character was an undeveloped skill. This matter was important because the meteorologic situation occurred in a few other places, the coast of southern Chile and the coast of North Africa. Because there was a possibility of an Allied invasion near Casablanca, Morocco, it would be highly useful to be able to forecast the daily weather of that place.

We worked under the immediate supervision of Professor Morris Neiburger, but under the general direction of the world's greatest meteorologist, Dr. Jacob Bjerknes. Lt. Charles Beer and I were given quite a free hand in devising different types of measurements and analyses that might precede new knowledge of the forces and patterns operating in the local atmosphere. Some of the tools had been previously developed by Professor Neiburger, especially time profiles of upper air wind and temperature. Beer and I took the matter much farther for we were able to get army personnel to assist in making simultaneous observations over a long distance. We also became deeply involved with the structure of the sea breeze, a subject that later I tackled in even more detail while I was in Hawai'i. We published several papers in meteorological journals and were among the first to see the encroaching problem of air pollution in the Los Angeles basin. We wrote an outline of a forecast method for stratus clouds that was less than what we had hoped for as a final result, but the end of the war cut the operation before all avenues had been explored.

During the several years this operation existed, Lt. Beer and I studied at night and on our days off. By this extra work and lots of study, we each earned a Master's degree in physics–meteorology. The time spent with Dr. Bjerknes, however brief, made a big impression on me. The great man spoke slowly and softly with the highest regard for his listener, and always with a respectful attitude, even to the most beginning of students. In the professional training program in meteorology, the most memorable day was when Dr. Bjerknes gave an unexpected lecture.

The cadets had just returned from the period of athletics, always exhausting, to where professor Bjerknes started his lecture in a darkened classroom. There was always a tendency to doze off under the circumstances, but on this day, I seem to have awakened, startled by a statement from the professor I had never heard before. The gist of it was as follows.

"As you can see from the combination of pressure and temperature, and the gradients of the two, the isobars are going to be highly compressed in that narrow band. Though never before measured, there must be a high-velocity wind in this zone, winding its way as a ribbon from one pressure cell to another." This was the forecast of existence of what we call the *jet stream*, now routinely measured.

Years later I was giving a lecture to the American Meteorological Society in Hawai'i, an occasion when Bjerknes was present. I told this story and said that this should have been called the *Bjerknes Wind*. He smiled and seemed pleased. Here was a case of pure deduction that forecast an important feature, rare in modern science.

Having completed my duty in the Army Air Force in 1946 I turned to a friend for advice on employment and took an assignment with the US Bureau of Reclamation in Washington. I did not much approve of the direction the Bureau was taking but I liked to use my training to increase their interest and knowledge of sediment in rivers. I designed and had constructed a fine hydraulic laboratory in the Denver office and used what influence I had to increase cooperation with the United States Geological Survey, which had carried out many field studies of erosion and deposition in the western states. In that work I was closely associated with Walter

B. Langbein and H. V. Peterson, through whom I learned about techniques of land management for erosion control, most of which were quite inadequate. I was not happy with the work being done by the Bureau because I did not approve of the large dams on rivers that they were designing.

In 1946 my father, Aldo, sent a manuscript of a book to Knopf, the publisher. The reply was something that sounded like "these are very nice sketches but the manuscript is too short so why don't you write some more and we will consider it." Aldo was furious, as one can imagine, for some of these essays had taken years of work before he was satisfied. When I heard this, I told my father that he was too soft hearted to deal with these commercial publishers so why not let me do it for him. Dad was delighted that I offered.

I began by talking to some of my conservation friends and obtained the information that Oxford Press had published some interesting environmental books. After several meetings in New York with the principal editor, we got down to the matter of the title, which was the only thing holding up their acceptance. Aldo had chosen the title, "Great Possessions," but the publisher thought it sounded too much like some 19th-century title, so I made up a long list of possible alternatives. The only one that seemed to appeal to them was "A Sand County Almanac," not my favorite but acceptable. We made an agreement and I was free to look for an illustrator. I was lucky to find that our friend Charlie Schwartz was willing to take it on so he and I spent some time seeking out sentences that we thought lent themselves to illustration.

Meteorology as a Profession

During this period, an acquaintance in the US Weather Bureau called me to see if I would talk to a visitor from Hawai'i, Dr. Eugene Auchter, who was concerned with problems of weather in the sugar and pineapple industries. Dr. Auchter explained that he wanted a long-range weather forecast of rain for Hawai'i. I told him I had no confidence that any long-range forecast of weather would be developed in the next few years, but I explained that a scientific analysis of meteorologic conditions could lead to a valuable short-range forecast system. After a few hours of talk he offered me the position of Chief Meteorologist in the Hawaiian Pineapple Institute and in the Experiment Station of the Hawaiian Sugar Planter's Association. I was glad to be able to find a different position and was delighted with the offer as well as apprehensive. I had never been out of the continental United States before. Before I accepted, I called my father to discuss it with him. As always, he was supportive, asked many questions, but never advised.

Auchter was as good as his word. Within a short time in 1946 we were sailing on the *Matsonia* for Honolulu. The trip to Hawai'i by ocean vessel was the first time I had been abroad. It was a thrilling experience. The first morning hundreds of gulls followed the wake of the ship landing to pick up food and other things thrown overboard. They were predominantly Herring gulls, and at the end of the day they had disappeared. Among the gulls were some black-footed albatross that never ventured as close to the boat as did the gulls, following a few hundred yards behind, often skimming close to the water. They seemed to keep up with the ship moving at 18 knots without flapping their wings at all. Some stayed with the ship all the way to the islands. I was entranced with the porpoises, the gulls, the sky, and the ocean.

At the pier in Honolulu we were greeted by Dr. Auchter and his wife, decked with leis, and taken on a local tour. The main thing we talked about was the local vegetation, for he wished me to recognize all the common species of tree, shrub, and flower. We were treated royally for two days and then I was taken to the simple building that would be my office and introduced to my new secretary, Jean, who I found was a swift and accurate typist and a sweet associate. I was given a car and a temporary place to live, the home of a professor on leave for a short assignment.

I busied myself with meeting the Weather Bureau personnel and Director of the Sugar Planters Experiment Station, Dr. Lyons, and getting acquainted with the geography. Some weeks went by and I had no instructions from the Director. I was quite concerned. Finally, there was a message to see the Director in his office. He greeted me kindly and then said, "I suppose you want to know what is expected of you."

"Yes sir, I have been very keenly interested to know what I am to do."

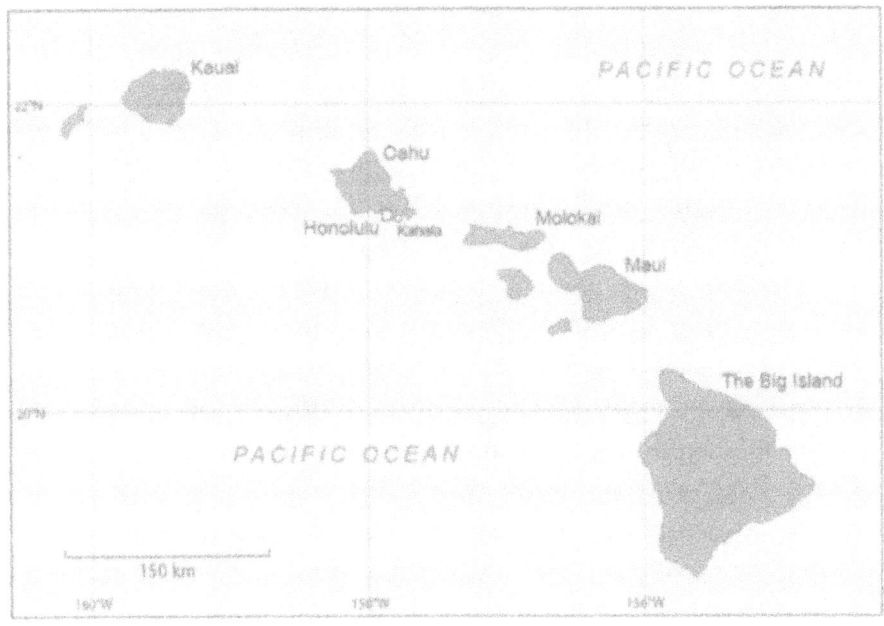

Hawai'i.

"Here is what I want you to do. Nothing." I gulped and awaited clarification.

"I want you to go all around the islands, meet the managers of the plantations, find out all there is to know about growing pineapple and sugarcane. Go to the fields and observe.

You have a nearly unlimited expense account, so please do what I say."

I was stunned, but when I thought about it, nothing could be more forward looking and practical for assuring the cooperation, interest, and dedication of a new employee; I never forgot those instructions, and I could see that freedom, support, and confidence were the cornerstone to getting the best performance from him.

Observing Natural History

To a new member of the staff of the Pineapple Research Institute it was immediately apparent that the work being pursued demanded a secure grasp of plant physiology, entomology, mechanical engineering, and soils. I was also a staff member of the Hawaiian Sugar Planters Association Experiment Station, which specialized in agronomy and plant physiology. Because my first assignment was to learn as much as possible about both sugar and pineapple culture and production, this joint appointment gave me access to the plantations on all the islands. Few others had this freedom and I was going to use it.

One of my first impressions of living near the shore was the difference in the sound of the surf. Off the beach at Kahala where I lived, there is a reef about 100 yards offshore over which the waves break. In the early evening the sound of the surf is loud but in the morning one can hardly hear it. This apparently is the result of near ground cooling causing the sound to be reflected back from the upper layers of warmer air.

In the zone of calm water behind the reef the native fishermen appeared in the evening carrying torches and short spears. In water waist-deep they watched for lua and squid. An incoming tide was considered best. Some used set lines baited with fresh shrimp about 3 inches long on a hook for bass or walleye. The shrimp was tied on the hook with heavy white thread. In the morning sunshine, local boys were out at low tide using sticks with a single barbed point and a wooden handle to poke into very small holes for squid. A glass box was used to see the bottom. The squid was not impaled but only irritated so he climbed up the spear and was quickly pulled off and put in a sack.

Kauai

On one of my early travels, I went with Dr. Nightingale, head of the Department of Plant Physiology, to Kauai where we were to talk to one of the plantation owners. One's impression on seeing Kauai for the first time is a series of deep, green, and dark valleys dissecting a high flat plateau edged with dark ridges topped by clouds. The sky was blue but broken by bulging shadowy clouds when dark shadows moved across the light-green plateau, flashing up and down the slopes of the valleys that reached back into myriad great hazy gulches between the mountains.

Driving east around the island toward Lihue, the road cuts through deep red soils where mosses and ferns cling to the shadowed road cuts damp with dew. The fresh moisture makes the soil appear unnaturally red and the great mango trees very black. The mango is a particularly beautiful tree whose smooth and shiny leaves, shaped in an unbroken ellipse, give a solid appearance of the tree, particularly in the morning light.

The road goes through groves of closely packed, rough-barked eucalyptus and through little gulches hemmed in by banana trees. The southern part of the islands is dominated by a ridge that juts out of the main mountain mass, the top of which is the Alakai Swamp, which runs northwest from the highest peak, Waialeale, on which 500 inches of rain falls each year. The general feeling of the island is of green and checkered fields on the spurs separated by sharp and jagged mountains or by narrow green gashes cultivated in little plots and small houses in a tropical atmosphere of slow speed and contentment.

At the headquarters of the Kauai Pineapple Company we met the supervisors, Mr. Watkins and Mr. Gregg, who took us to see their fields. Dr. Nightingale explained his method of judging the plant's requirement for fertilizers. The need for nitrogen can be determined by the color of the pineapple leaves, a method he developed. In plants deficient in N, the leaves begin to turn yellow instead of a good green. He developed a color chart for comparison in color. The color called No. 1 is when the center line of the older or basal leaves takes on a reddish tinge, usually flanked on each edge by normal green. No. 1 color means a deficiency of nitrogen. If the nitrogen is added two months before differentiation, the past deficiency will not hold the plant back. Differentiation is the first appearance of a fruit bud, which sets in the center whorl of leaves; it is noticed by a reddish color appearing in the very young leaves deep in the central whorl. After the reddish color appears, the young bud becomes a large terminal bud. The color of reddish brown is surrounded by bright red on the surrounding young leaves. As the bud enlarges and grows to be an inch or so in diameter, long purple flowers appear protruding from each eye of the growing fruit. Practically no crown leaves have yet appeared on the fruit. The fruit develops a crown and enlarges slowly.

Planting is done using crowns, suckers, or slips. The sucker is a vegetative branch coming off the main stem near the ground. The slip is a smaller vegetative branch appearing immediately below the fruit. In planting, long strips of black plastic are stretched in rows across the prepared ground. About every 2 feet a hole is plunged in the plastic and a short piece of planting material is pressed into the soil through the hole in the plastic. The plastic inhibits moisture loss and keeps weeds from growing near the plant.

On this trip we visited another plantation, Grove Farms, spending a morning with Mr. Will Alexander, the manager. This farm is primarily a sugar operation but at that time, they were just starting into pineapple, to which they had devoted a couple of hundred acres. With considerable effort, this plantation constructed a mixing plant where fertilizers and spray emulsions are mixed very rapidly and efficiently. They have the good fortune to have on their part of the island plenty of water and thus can apply 1,000 gallons of liquid per acre. This amount is considerably more than could be used on the dry islands, where the same fertilizer or sprays are applied with 400 gallons or less. A boom spray in a field of pineapple travels only 800 feet on a tank of 1,000 gallons. The spray boom is 50 feet long so they hook a slow truck to the spray tractor and pump into the spray tank from the truck as the boom machine proceeds along the field. The fields are generally 100-feet wide so that a spray boom can reach halfway across the field from each side from the road along the border.

In fields where sugar was planted following pineapple, the cane grew exceptionally well, but the operators fear that the cane will be low in sugar content as a result of the luxurious growth: Low in "bricks" they would say.

On another day we went to the cannery of Hawaiian Fruit Packers Ltd. and talked to the manager, Mr. Dorsey Edwards. This outfit started only recently and did not have access to very good planting material, only Cayenne crowns that were mostly "collar of slips." This means a large number of slips came from a collar at the base of the fruit, a rather undesirable trait in that much of the growth that should go into the fruit is expended on a large number of slips.

In this plantation there was a considerable wilt in the fields and some yellow spot. Particularly in the south, on the Kalepa Ridge, much of the landscape is grassy or at least bare of forest cover on the makai side and forested on the mauka side, which is also the windward side.

When I was on Kauai, I had the good fortune of meeting Charlie Schwartz, the great naturalist-artist whom I had recruited to make the drawings for *A Sand County Almanac*. He was in Hawai'i on an extended trip for photography and a study of the wildlife. He took me on my first trip to the northern tip of the Kauai highlands, Kokee. We went up the valleys of Hanipepe and Waimea as far as the roads went on the valley floor. These valleys had numerous patches of individual farms where rice, taro, and some beans are grown. They are lush and beautiful, bounded by high cliffs with red soil and black rocks.

The road to Kokee goes up the west side of Wiamea Canyon. Though we saw no Hawaiian ducks there were a dozen coots, which are a little smaller than those on the mainland. We saw one gallinule and a number of golden plover. On the road to the northern crest we saw numerous apapane, one of the few remaining Hawaiian species; a sparrow-size bird, deep crimson in color with black wings and tail. They were feeding on the red torch of the ohia lehua tree. We hoped to see the rare iiwi.

From the summit one looks down the nearly vertical cliff of the pali to the waves breaking on the shore below. It is a breathtaking sight. We did see a few of the jungle fowl, ancestor of the domestic chicken, probably introduced by early Polynesians or possibly by the Chinese.

A fine mist filled the air. The trail was slightly muddy and the drops of moisture continued to fall from the leaves of the ohia trees. A few of the large ferns stuck their new, lettuce-colored leaves out into the trail where the sunshine, filtering through occasionally, helped to unfold the minutely articulated leaflets. We were high now, I suppose nearly four thousand feet, and for the most part above the clouds, but there was still sufficient stratus above us to continue the fine drizzle.

As we walked, a gradual increase of wind told me that we were nearing the edge of the pali. A moment later, there was an abrupt break in the trees and I was facing the horizontal line on which ferns, scrub ohia, and a wet grass were silhouetted against the unbroken gray of the fog, actually the upper part of the cloud. Since I did not know how steep the cliff was, I walked the last few yards gingerly, and holding on to a small ohia, looked over the edge. To my disappointment everything in front of me was the same gray of the fog. Though I could see the that the edge in front of me dropped off very steeply, the last visible thing below me was a native palm that was hanging onto the cliff about twenty yards below. This native palm is the only one that occurred in the Hawaiian Islands before the introduction of the coconut by the early Polynesians. In this mostly tropical forest, it seemed strange that a native palm should occur only on the windswept cliffs covered with greenery, mostly staghorn fern and a few shrubs, but there was no exposed rock even in the steepest section.

Suddenly there was a small break in the cloud cover to my right, the fog had parted near the cliff and I could see down into a bottomless hole, bounded on three sides by walls of the writhing vapor, the fourth side being a vertical wall of ferns and palms, all getting smaller deep in the hole.

While I was still looking for the bottom, there was a noticeable brightening of the whole atmosphere, as if the sun were getting ready to melt out the rest of the cloud. In a moment another break in the fog layer opened a new and much larger vista out to the left and for the first time I saw the bottom of the pali, four thousand feet below, a smooth curve of treetops receding away from the base of the cliff like the surface of a green wave off Makapuu Point. Interstream divides separating the small watersheds at the base of the cliff swept away from the base in little pyramids that seemed to support the wall and keep it from falling over. In between these was a smooth surface in which light ribbons of kukui could have been molten bronze flowing down the draw.

In the meantime, larger sections of the cloud had dissolved, until now the illusion of a hole in the fog was destroyed. The whole expanse of the cliff became a reality and the remnants of cloud floated below, casting bent shadows as they passed over alternate hill and valley. The larger of the interstream divides now appeared as good-sized ridges, also bounded by steep slopes and generally culminating in knife-edges, which stuck out of the main cliff face like scimitars plunged in the pali after a duel. The hilts and guards of the swords were crags that protruded out into the ocean. From the curve of sandy beach marking the mouth of one of the valleys draining the pali area, the color of the ocean darkened with increasing depth, pale green near the shores, mottled green and violet inside the reef. Beyond the reef, where miniature breakers formed a moving line of foam, the color suddenly broke into indigo in the sun, and Prussian blue under a cloud. In one of the valleys below, I could see a light-colored patch marking the extent of a bamboo thicket.

I was looking down on a number of hills or peaks of considerable size but there was no yardstick by which I could get a true measure of their size. A long section of the pali could be seen stretching away to the place it made the break in the horizon profile. Looking along the direction of the pali, the rugged nature of the summit was evident. At the high points the pali was cliff on both sides, extending as a knife-edge into the sky in the same manner as the interstream divides over which I was looking. The windward side of the edge was somewhat steeper than the lee and the vegetation on the exposed face much smaller and more herbaceous in character. The lee side was thick with trees, even on the very steep portions. These trees, mostly ohia, clung with vertical roots to the cliff face and after bending away from the face of the pali to reach the sun, grew upright or nearly so. The tops of these trees grew as high as the top of the knife ridges itself but at that height were trimmed off and shaped by the strong winds ordinarily hooting up the pali face.

This combination of the relatively bare face, the knife-like crest, and the wind- shaped treetops on the lee side gave each peak the appearance of a breaking wave, while the tree tops leaning out to leeward were the foam and spray blowing off the wave crest.

I watched the clouds drifting across the base to the pali and the patterns of the shadows on the ridges. Since both clouds and shadows were below me the shadows seemed to race up and down the hillslopes trying to keep up with the fluffy carefree wisps of cloud.

While I was watching the clouds, a small white object attracted my attention as it was floating below. It disappeared the moment I saw it and while I watched the point waiting for a cloud to go by, another white object came in sight, this time much nearer, for a moment it was just a white triangle, but then as my eyes shifted into focus, its birdlike character was immediately obvious. The bird never moved a feather but rose swiftly until he was just below me and near the cliff. With great speed he swept up past the crest of the pali, which pointed up at the graceful curve of his wing. The slim white needle was the incandescent trail of jet-propelled motion. The white-tailed tropicbird, riding the surf of the pali wind, up from the mortal ground, over the sheerness of the abyss, then high into the Hawaiian sky. What more subtle way to emphasize the immobility of this poor stranger to the tropical rainforest. Not a ruffle of a feather, not a wing stroke, up through a cloud and over the top of the pali.

I looked down a to see if there were others. Just below the cloud were three more, gliding effortlessly, spinning great circles for the pleasure of hearing the wind or perhaps for the fun of watching their shadows climb and dive over the ridges buttressing the pali. Suddenly I saw two more farther down, and more down still closer to the cloud shadows, not just one or two, but dozens. One finally passed close by me. I then could see the yellow beak, the black stripe through the eye, and etched on the pure white body the bold patches of black on the top surface of wing and back. But most striking was the narrow tail that projects out behind like a shining dagger pressing him forward.

I realized that I had not seen the birds because the magnitude of this pali had not impinged itself on my consciousness. Now the sight of those specks of white moving smoothly and effortlessly under the clouds began to convey the idea of the height and extent of the cliff.

Just as quickly as the fog had cleared with no more apparent reason, the clouds below were coalescing and in a few moments the bottom of the pali was banked with fog, and only one bottomless hole remained in the swirling vapor. Then it too was filled and the abyss was merely something dreamed. Were it not for the recollection of the white bird with the long spike of a tail, the whole scene could have been just a thought. Only a tropicbird can be an adequate guide through the showcases of the Hawaiian morning or in the intermittent view of the rainforest.

Local Fishing

On a day when the tide was exceptionally low, I waded out to the reef that was 150 yards offshore of the beach near where I lived. The variety and colors of the reef fishes were extraordinary, some silver with small black dots, black and deep with orange stripe behind the gill, a number of Moorish fish, yellow and black with long streamers from the dorsal fin; one with a stripe of gold and both dorsal and ventral patches a beautiful bright blue.

On another occasion I met a young Asian man, Herbert, who showed me something about fishing. He used a very small hook, and a whippy bamboo pole about 7 feet long baited with shrimp. If he got even the smallest bite he would yank like hell, before the fish sought refuge in a puka (hole) in the coral. He got a few small pan fish and a strike from a real fish, but the thing got into a hole and the line fouled. He judged the fish was a papio, a beautiful prize when you got one. To bait the hook Herbert just mashed off a piece of shrimp with his thumb, holding the shrimp in one hand the hook in the other he tied it on to the hook with a short length of string.

Agriculture

I learned much about growing various plants. In his garden a friend showed me the roots of green beans that were covered with nodules of nematodes, a class of worm that attacks many plants. About the time I came to Hawai'i there was an outbreak of some insect attacking pineapple plants, a serious infestation, the reason for which was unknown. The head of the Department of Entomology, Dr. Walter Carter, started a study that looked at first as if it was going to take a very long effort.

Carter found the fields were being destroyed by some kind of insect. Some plants had mealy bugs on the leaves and they also harbored ants but there seemed to be no connection between them. The mealy bug of the Homoptera group is a sap-sucking insect. It was clear even at the beginning of the investigation that the dying plants had no mealy bugs. Observation and experimentation led Walter to conclude that the mealy bugs were infecting the plants and that ants were milking the mealy bugs of their excrement, which contains sugar and amino acids. The ants drink this and carry it to their young. But why there were none on dying plants remained a mystery. He found that the ants were carrying the mealy bugs to fresh plants when the infection was affecting the growing plant. The ants were apparently herding the insects, like cattle that are moved from one pasture to another. Once this discovery was made, known pesticides were used to control the outbreak. Walter had, for a second time, saved the industry. This was an example of how the two research organizations for which I worked served the industries in a fruitful way.

Molokai Landscapes

Another example took place during the time I was working in Hawai'i. The Sugar Planters Experiment Station was also exploring the possibility of commercial production of various fruits. Dr. Lyon, the Director, took me to Molokai to look at the field substation where, among other things, he was personally experimenting with varieties of mangoes. There were at least 20 fruit-producing trees, all having different color, size, taste, and plumpness that he displayed. That was a most interesting exhibit. On another occasion I went to Molokai to meet some scientists of the Experiment Station, landing at the airport west of Kaunakakai that has a rather short runway in the N–S direction.

Molokai is an island shaped like a rectangle about 37 miles in the E–W direction and 10 miles in the N–S. It is divided about in half along its E–W axis, the northern segment of which is a rugged mountain bordered by an impressive nearly vertical cliff on the north some thousands of feet high. The southern segment is a low-lying nearly flat plain covered for the most part with alluvium on which early people could profitably grow crops.

I have long thought about the origin of these two landscapes. My best conjecture is that the volcanic eruption that produced it was a line of several vents along an E–W axis. The north border faces the trade wind and the violent surf that batters the base of the cliffs that must have been formed by continual erosion. The original shape of the north has been eliminated as the cliffs retreated and the erosion products built up the south shore. Thus, the south half on the lee side of the island must have been the recipient of erosion products from the adjoining mountain mass to the north. On this deposit is the town of Kaunakakai that had two streets and a rather lazy appearance, reminding me of old western towns with false fronts. There were, however, a surprising number of stores, four groceries and a couple of general stores. A long wharf extends a mile into the reef. A channel had been dredged along the end of the wharf for sugar and pineapple barges to land.

The reef on the south edge of the long island is covered with soil material and there are few sand beaches. Much of the shore is muddy with clay that makes it dark in color. There is a large number of ancient fishponds built and used by the native Hawaiian centuries ago. This relatively flat area of alluvial soil making up the south half of the island is larger than any other in the island chain and though I do not have the data to show it but I conjecture that there was a concentration of native population here because of the area available for agriculture.

Even in my brief experience in the islands, I saw more ancient fish ponds along the south shore of Molokai than any other place. As I understand from several natives, these fishponds were built by placing rocks in loose walls surrounding an area of water. At the time they were built the mud that now dominates the enclosed area was absent. The loose character of the rock walls permitted the easy in and outflow of the tide and fish could swim through holes between rocks. The normal distribution of fish meant that a reasonable number could be within the pond at any time. A native fisherman walked out into the pond, the fish would scatter, but they could not swiftly find holes in the walls so they were easily speared. On one of our trips to the experimental plots on Molokai, Hugh Brodie and I took time to look for the various structures and other indications of the early inhabitants.

The shore along the southeastern border of the island is rocky with only a few isolated beaches, most of which have black sand from the basaltic rocks. The water off this shore was exceptionally blue. From there the road (a simple track) sweeps north, where it enters a grazing country. The grass seems fairly good but old erosion scars on the slopes indicate heavy use on the Fagan ranch. Beyond the ranch headquarters the road comes out on the edge of a plateau that looks down into the Halawa Valley that resembles the Hanalei on Kauai but is narrower. There are remains of old taro patches but much taro was destroyed by a recent tsunami.

The upper part of Molokai Ranch country can be reached from Mr. Cooke's ranch, to which we got the keys from the caretaker. Leaving the main road that connects Kaunakakai to the airport, we went north through a section of red laterite soils and poor grass. Water for stock is carried by pipes to wooden storage tanks from which it is fed to a watering trough through an automatic control gate. This water is obtained from high-elevation streams or springs.

From the red soils one comes abruptly to areas of brown or purple soils. The lower grazing land is scattered with koa trees to a zone devoid of trees above which is the mixed ohia forest. There are large patches of planted eucalyptus in the ohia belt. In the rain forest there are isolated houses one of which provided shelter for a Japanese caretaker whose job apparently was to maintain the water supply pipes. Some miles up the mountain the road became so slippery that we proceeded on foot. Rain was intermittent through the ohia forest, the floor of which was covered with bracken fern and patches of sphagnum moss.

I left the trail and went toward a place devoid of trees that I suspected was the top of a cliff. A mist blew up over the edge of the pali and when there was a slight clearing, I could make out the other side of the valley on whose edge I stood. It was about 200 yards away, perfectly sheer, but swathed in a luxuriant growth of ferns, small ohia trees, native palms, and lauhala. There were a series of waterfalls, some of which had water and spray. Each one waterfall was a vertical trough devoid of vegetation that provided a vertical brown line contrasted to the adjacent green cliff. Each terminated in a plunge pool some hundreds of feet below that overflowed so it was followed by another free fall of a thousand or so feet, followed again by a repetition of the one above.

There was a brief clearing and suddenly I could see the whole valley at one time and I could hear the muffled drone of the distant surf as if on a model, a miniature curving beach. Looking down I could only see a few hundred feet of cliff because the rest was hidden by an overhang. In a moment I was looking into a dense mist and little spots on my eyeball appeared to roam about because there was nothing on which the eye could focus, just brilliantly luminous fog. It was as if I had imagined the whole thing, a thought that struck one forcibly here.

The rain began in earnest and we turned down the trail. We traveled westward to a place where there is a trail leading down the cliff to the leper colony situated on a small, low peninsula extending out from the base of the cliff. The trail was blocked off with a gate and barbed wire. That is a shame because a physician I know in Honolulu is the principal doctor serving the patients there and he assures me that there is no danger of getting leprosy from ordinary contact with those who suffer. All the fear is a leftover of centuries of myth from premedical times.

Not far from this location we found the large phallic stones that old Hawaiians thought were helpful in cases where a wahine (woman) was barren. She would go up to the stones and sleep there on the theory it would make her conceive. The stones are points of basalt that stick up out of the ground at odd angles.

At lower elevations outside the rain forest there are other objects of great age built when the population was large, and before these talented people were devastated by the diseases brought in by colonists and were subjected to other religions considered more truthful or more valid than their own. These people built elaborate platforms standing about 10 feet above the natural surfaces with a smooth flat surface about 75 feet on a side. These rock structures have survived nearly intact for centuries. The local people say that the platforms were used as assembly points where the religious ceremonies were held and probably for other purposes when a gathering of people was desired.

In this general area we came across another old and equally anachronistic custom where it was played out. There was a flat open area about an acre in size on which an elderly man was growing rice. One of the problems with this crop is that it can be nearly devoured by a flock of ricebirds. This little brown bird, considerably smaller than a sparrow, congregates in large flocks.

The man was reclining on a wooden platform he had built about 8 feet above ground. Atop this perch a vertical post extended another 3 feet and from this post were more than a dozen strings, each stretching across the field where each was tied to a short vertical rod made of a branch of a local tree, the rods spaced around the field at intervals of perhaps 10 feet. The full length of each string was flying small white flags about 3 feet apart.

We watched a flock of about 100 ricebirds settle in over the whole field and begin to feed off the rice plants. The man who was drowsing would awaken, sit up, grab the place where all the strings converged above his head and shake the strings as stridently as he could, the little white flags all fluttered at the same time, the flock of birds would rise simultaneously and be off in a dark cloud of motion. The man would resume his nap but within five minutes the flock came back, settled into the crop, and the man again shook the strings, the flags fluttered, birds departed. It was genuinely comical, but after this repetition a few times, when the birds had settled, he picked up an ancient shotgun that he had on the platform and shot it off in the far distance with a very loud bang, the flock rose again, and didn't return for about 15 minutes. The whole drama was a once-in-a-lifetime experience. Whether he saved any of his crop is lost in the fog of history.

Crabs, Fish, and Research

On the beach I watched some sand crabs building their holes so commonly seen on all the shore. They are wary creatures so one sees lots of holes but no crabs. The animals spied me 150 feet away and dodged for water or for a rock pile. Once they are in the water they dig rapidly until only their back shows and even then are well camouflaged. Evidently, they go only deep enough to allow their protruding eyes to survey the situation. I sneaked around and projected only my head around a stone wall at the edge of the wharf. I saw one just coming out of a new hole he was building and he stopped short, froze, having easily spotted my head. For about 5 minutes he sat motionless, but because I did not move, he finally wiggled his eyes to see everything, and haltingly moved toward the hole. Just before disappearing, he waited again to look. Finally reassured, he went in the hole and in about 10 or 15 seconds emerged carrying a load of dirt between his big claw and three or four of his feet. He carried it smoothly over to the sandpile and with a real throwing motion tossed it up on the pile, at the same time reversing direction of travel and returning to the hole in reverse gear.

The whole operation was smooth and efficient. He generally builds the dump pile seaward of the hole, presumably so it will be a haul in the downhill direction. Watching this show for some time, I finally jumped toward the hole to see what he would do. He came out as usual until he was full in the open and seeing me, headed in reverse shift as fast as one could imagine. He did not come out again.

An elderly kanaka (native) was fishing on the reef. He used a very small hook and baby shrimp for bait, opal in color and about $3/4$-inch long. These shrimp occur in very shallow water on the muddy shores. I watched him catch a 6-inch papio, of beautiful shape, all silver with a fine blue line along the body. It was a real fighter.

The party I met consisted of Dr. Mangelsdorf, geneticist; Martin, pathologist; Williams, entomologist; and the resident caretaker of the station, George Otsuka. We were to inspect the quarantine area where new varieties of sugar cane are being grown and tested. It consisted of an outdoor field of planted specimens and a greenhouse that is protected from any contamination. The samples of cane received from the quarantine station in Washington, D.C., are first planted in sterile pots containing soil. In the greenhouse the plants are watered each day and grown for a year. Complete isolation is achieved to prevent any new disease from contaminating the plant. Only a few people go into the greenhouse. They wipe their feet in creosote, go to an anteroom where they change into sterile clothes, knives or a hand lens are dipped in Lysol. Only then does one enter a dark room, closes the door, all before entering the quarantine room.

We cut some field cane, the stalk numbered, inspected, and given a treatment in hot water. About 5 pieces about 2 feet long were to be sent by air to field planting stations at Hilo, Kailua, and Kauai for another growing period. If at any time a symptom or disease is noted, all the cane in the planting is destroyed. A water treatment of 52°C for 20 minutes will kill any spores or insects not visible.

The resident caretaker of the station, George Otsuka, is short and sinewy, about 49, and worth listening to. While he was cutting cane, he gave me an account of cane knives. Sugar cane cut by hand is an operation of skill. The fibrous stalk must be cut with a sound blow and a good knife. Those made prewar are much better in quality than those presently available. The old knives were made of better steel and of better design. The last couple of inches of the blade are very important and should be the widest part of the blade. The new knives, he says, have a belly bulge, are too wide in the central part of the length. The old knives have little or no bend but are stiff, and have a handle riveted on with rivet and washer. The new ones have a single pressed rivet mashed at the end.

As a special treat for me he took me out to one of the ancient fishponds consisting of a loose rock wall enclosing an acre or two of water, inside the coral reef. No Hawaiian remembers having heard a date of their construction. These early people did a lot of this stonework, the remnants of which are in the heiaus (religious platforms) and the fishponds. The walls are loosely constructed, triangular or trapezoidal in cross-section, readily permeable to the tides. They are usually more filled with silt than water, filled from soil washed off the hills.

George waded out about 30 or 40 feet from the wall at the makai (seaward) end of the pond, laid out his net parallel to the wall. The water was about 3-feet deep and mixed with mud. He discussed at length the trouble when a Samoan crab grabs the shin or calf with its large strong claws. He said if it grabs you, you must stand still despite the fact that the claw can sink in your flesh an inch or two. In a moment the crab will let go and swim off. If you yell or jump, the claw will tear your flesh and you are liable to break the claw off the body of the crab. In this case its muscle will stay contracted, and you will have a hell of a time getting it released. Even a heavy pair of pincers will have difficulty cutting the shell of the claw. The other way is to press the claw deeper into the flesh, which will provide some slack in the muscle and the claw can be forced open. The claw of a Samoan crab is a mean tool, protected by a tough, thick, bony shell. These crabs lie in the silt during the day but at night come out to scramble around the shore or the rock wall in search of food.

In moving around, the crab may hit the net of the fisherman and get entangled. When the net is pulled up and has a crab, the fisher grabs the crab by the two claws, one in each hand, and breaks the claw off. This is obviously tricky, because the claw must be grabbed in such a way as to avoid being pinched. Moreover, one must not grab it in the angle of the arm near the body shell because it can crush you inside the elbow.

He instructed me to follow him through the ancient fish point to the reef but told me to step only in the places his foot had been put in the mud. I followed him as instructed and out on the reef he showed me the art of throwing a fish net, a woven net of about 10 feet in diameter that when thrown correctly, settles in a round circle that will trap any fish in that area. It surely took years of practice to attain the level of perfection that George demonstrated.

George brought up 6 or 8 crabs that were all alive but without claws. They were dark green in color, the carapaces being about 8 or 9 inches across. He boiled them about 30 minutes in salted water, starting them in cold water. We ate crab that night.

Views from the Coastal Cliffs

On another occasion, Gordon Nightingale invited me to accompany him on a hike along the crest of the Koolau Range. We left Kahala about 7:30 and drove through Wahiawa to a pineapple field where there is a primitive road leading to Holemano Camp. I had seen lightning before dawn and expected considerable rain but we were in a moderate Kona condition where there is a tendency for a storm from the southeast. There was a high ceiling of middle clouds and wisps of stratus over the mountaintops. John Fo from Gordon's office drove us as far as a jeep would go, a couple of miles beyond Holemanu.

All of the roads and trails in these mountains follow ridge lines of the mountain top. Where we started to walk, we were already high enough to see both koa and ohio lehua, an elevation slightly above 1,500 feet. The narrow trail had been built by the C.C.C. during the depression years of the 1920's. It is cut into rock over considerable distance. The rock is deeply weathered and, in some places, can be gouged with a pointed stick; it is still a result of great effort that now affords a visitor to see the high country. Even on this type of day with no rain, every leaf appears wet with dew or recent cloud fog.

The trail is partly or in places completely covered with vegetation, drooping from the adjacent hillslope or grown up from the downslope side. Such blocking of the trail did not impede walking. This ground cover consisted of staghorn fern, ti, guava, and a large amount of red raspberry that is bright red, but individual fruitlets are very small so that the outer appearance is more like a strawberry. They are insipid in taste though they look delicious.

Gordon and I walked 5 or 6 miles to get to the crest but the rainforest was a delight. We saw several apapane and I caught a glimpse of a bright-red bird with black wings, I was sure it was the rare iiwi though I did not see the long, curved bill.

Near the crest we came to an old house near which was a rain gage of local manufacture consisting of a regular gage 30 miles in height and 8 miles diameter but it was connected to an overflow rubber hose that led to a separate storage tank. The regular cylinder was full and the auxiliary storage tank partly full. I do not know how often it is visited. The whole rig was held upright against the wind by a flimsy array of baling wires attached to stakes.

At the crest, that on this day was free of clouds, we could look down into the deep gulches and out to the beach. Kahana Bay lay a little south of us and beyond the Piu Manamena we could look into Kaneohe Bay to the Naval Air Station. As you first look down you saw a nearly vertical pali some 1,500 feet high, the sides of which were covered with ferns, some mosses, lauhala and a native fan-shaped palm in the crevices. As your eye took in the tremendous length of this vertical cliff you began to get the impression of the depth to the more gentle slopes at the foot of the cliff, where a forest appeared as a carpet of green grass. Even the slopes of the intervalley spurs were very steep—but still forested. The native palms at this elevation were a surprise and a gratifying note. Each leaf was fan-like in shape with individual leaflets connected to one another nearly out to the end. The cross-section was therefore zigzag as each leaflet was troughed.

On the windward pali as we see from the crest, the near vertical cliff is partly covered with large ferns, occasional banana, and strawberry guava. On the leeward side the ohia is the most prominent tree and its rough bark is mostly covered with moss on the lee side of the tree. Its leaves can be of different forms, small or large, light or dark green, smooth and shiny or dull. We looked for but saw no sandalwood.

There was practically no wind even on the crest because of the Kona that day. Normally a hurricane trade would be found to nearly blow you off the edge. The trail was cut into the rock of the cliff, generally 50 or so feet below the crest, but often crossing over a knife-edge saddle to the heavily vegetated leeward side. For the most part the trail is on the windward. On the steepest cliffs there was generally no vegetation along the trail but where the slope was 60° or so. The outside edge of the trail was lined with ferns, grass, and herbs, giving a sort of guardrail and an all-too-false sense of security. The most surprising part of this peak topography was the great distances of steep cliffs and the steepness of the lee as well as the windward side. On the lee side even on steep places, there were many trees and each narrow crest gave the appearance of a breaking wave because the blown trees on the lee leaned over and appeared sculpted by the wind.

The trail was about 3-feet wide but in some of the steepest parts 4 to 5 feet. In places, however, it was a scant 2 feet and very uncomfortable. The height and sheerness bothered me for the first mile but I felt better after that. For the whole distance I was severely conscious of the abyss and I watched the trail intently, looking over the edge only when I stopped.

The distance along the ridge to the heads of Waikane and Lipapa trails is about 3 1/2 miles. There is a short spur to what appears to be a summit and there is a concrete pillbox where a machine gun crew was kept during the early part of the war. The Waikane trail is much steeper with much more vegetation in the path, apparently not well traveled. It goes down fast into heavy timber where there is a road leading to a water tunnel and house at an elevation of about 1,000 feet.

It rained as we made our way down that trail and was quite slippery. Nightingale said that rubber or composition soles are better than my hobnailed boots when any mud is encountered. A wide-brimmed hat, short jacket, and woolen shirt are practical. A walking stick is in my opinion very important. Though the whole area is very wet, there is water available for a drink only at a few places.

When I lived across the street from the beach, I learned much about the reef and the near-shore quiet zone. Once when I was fishing on the reef, I learned that the hooks I was using were much too large. The local kanakas used a hook about the size of a trout fly hook that would keep fish as small as 4 to 5 inches, like a small trout. I caught a few of these but threw them back, but they were beautiful to see. I saw larger fish, 6 to 8 inches, swimming around me but these paid no attention to my bait. I saw one man who had kept many small fish, including the humahuma nukanuka apu aa.

I saw one local man spearing fish on the reef. He had two large gunnysacks full of lobster and fish, including a beautiful papio. He told me to do my casting in places where a light color of the ocean indicated a sandy rather than a rocky bottom. On that occasion I had two fine bites but again I think my hook was too large. After I put on a smaller one, I didn't get any strikes but lost my bait to the reef fish.

While I was fishing on the reef with the small hook, I realized I still could not seem to feel these small fish quickly enough. I see why the kanakas use a very slender and rather short bamboo pole. Some local children were fishing with little 2-foot sticks and only 3 feet of line. They had caught a string of little fish 3 to 4 inches long. Among those caught were a manini of the silver variety with about 4 vertical black stripes. They caught a balloon fish that they gingerly threw off the hook without touching it. That fish is poisonous to eat and has a solid-bone jaw containing welded teeth and a good mouth to keep your fingers out of.

A Chance Observation That Changed the Direction of My Research

I was in a light plane flying over Lanai to one of my study sites. The wind vane at the NE end of the strip caught my attention and the wind was clearly from the NE. But the opposite end of the runway also had a wind vane and, to my surprise, the wind was from the SW. The NE trade wind was meeting the onshore sea breeze. Indeed, I saw that directly above the position where the two met there was a narrow line of clouds extending the full length of the island.

That one simple observation changed a portion of my research agenda. I found that there was little in the scientific literature on the subject, so thenceforth it took up an important part of my effort. This was to be only the first of just such small unexpected observations that changed my direction. I have no doubt that others have had similar experience being reoriented by some simple unforeseen fact.

I began a serious inquiry into how the sea breeze and the trade wind interacted on various islands, each of which has mountains of different size and orientation. The cloud line on Lanai was the most obvious response merely because the island is narrow in the NS direction but stretches EW for several miles. Maui is quite different because the great mountain of Haleakala extends well above the inversion and above the sea breeze. So also is the big island, Hawai'i, which has a peculiar local variation of the interaction zone.

The same phenomenon explains why on the N or NE coast of Oahu, Lanai, and Molokai, the usual scattered clouds typical of the nearby ocean are quite absent for a few miles away from the N shore. This is the zone where the sea breeze reinforces the trade wind leading to subsidence and warming so the clouds do not form along the N coast.

Ranching on the Big Island

My friend Geoffrey Davies is the scion of the Theo. H. Davies family and he was learning the internal workings of this far-flung organization engaged in sugar production, shipping, cattle ranching, and investments. He occupied the family mansion in the Manoa Valley of Honolulu and was a frequent user of the company's ranch property northwest of Hilo on the windward side of the big island. On more than one occasion my family and I were invited there as guests. The Kukaiau Ranch was in the center of the district of that same name in a little grazing country on the slopes of Mauna Kea.

The lower part of the ranch is in the ohia lehua forest uphill of the sugar plantations. The uphill edge of the ohia changes abruptly into the Koa forest and gradually shifts to the mamani forest at higher elevation. In the Kukaiau area there is a rather wide band just below the Koa forest that is nearly devoid of trees. There was evidence this was once a Koa forest that had died off. Single trees of considerable size remained standing among many dead trunks. The local corporal told me that this change took place in the early part of the 20th century. It was his recollection that there was an invasion of what appeared to be army worms. There was also a forest fire in 1902. These two apparently caused the belt of no trees and I saw fire scars on some trees. The local forester said that such an invasion of worms was periodic. Though he doubted that defoliation of the trees by the worms could account for the belt of deforestation.

I saw a remarkable lack of the reproduction of Koa everywhere I traveled in the islands. Young Koa is very palatable to cattle and thus there is no rest for the vegetation so Koa cannot get started. Dr. Lyon told me that Koa will reseed naturally if relieved of grazing pressure and he cited instances of reproduction in quantity on Oahu in areas that were plowed but protected from cattle. Both Bryan, the forester, and Dr. Lyon agreed that ohia needs to start germination while supported in the air by something, a tree fern often furnishing the proper habitat. Yet I noted that other reproduction was common within some paddocks of Kukaiau Ranch. The reproduction that I saw was mostly on the banks of gullies that had nearly vertical sides to erosional cuts. Yet in the central and the Kohala (north) area of the ranch I recall no ohia regeneration. As for the ohia germinating on a tree fern I saw many large- and medium-sized trees that had multiple root branches in the air as if the roots had grown down from a tree fern stalk.

The greatest gullying I saw on the Mauna Kea slopes was in the vicinity of the ranch headquarters as is the case in many ranches.

Cattle trailing seemed more important than just grazing in starting gully erosion.

Nearly all the grasses that make up the ground cover under individual trees in the Kukaiau region were introduced or exotic species. There were no grasses as I could see that represented the original cover conditions. The whole aspect of the country was that nearly all the ground surface covered with grasses was under trees. The dominant species I am told were sweet vernal, rattail, Kikuyu and Dallas Grass. According to Frank Corriea that grass was introduced in the late 1800's. All that introduction was for the purpose of range improvement.

On the Parker Ranch on the saddle north of the Kukaiau area, I did not see any of the grazing land in as good condition as at Kukaian. Parker Ranch runs some thousands of sheep without herders, a practice common to all the Mauna Kea area. The herds and the wild sheep and goats have eaten down all the Hale Pohaku area. On Kukaiau Ranch there were 5,000 head at a density of 7 acres per head year long. Parker has a lot of desertic area where but little forage occurs.

At Hale Pohaku the only grass is under the manaui trees that are widely spaced. In the extensive areas between manaui trees there is little or no grass and it consists of Pili (mountain grass), not very palatable for stock. The grass under the trees is supported by fog drip and the somewhat reduced evaporation under the shade.

One day I was exploring the ranch on horseback, in the forest zone of Koa I came across three men working on a very large Koa log that they had felled. They were carving it into a native canoe. The hull of the craft is made completely out of one log except for the vertical, beautifully carved posts at bow and stern. The artisans had outlined the shape, and at the stage I saw it, were carving the sidewalls with an adze. The hollow of the canoe was being developed by a small open fire kept burning with small additions of twigs and chips. After a period of careful burning the sidewalls were carved by hand, gradually thinning the shell. This log was about 30 feet long, the largest Koa I had ever seen and undoubtedly very rare in a Koa forest. The canoe is made very deep and narrow, the sidewalls amazingly thin and uniform in thickness. With the deep and narrow hull, stability is maintained by the outriggers.

In the years I lived in the islands, I never saw a new canoe nor did I ever hear of anyone who had seen one, so my experience was unique and rare. It would not be a surprise if there were no standing koa tree left in Hawai'i suitable for a canoe and probably very few craftsmen who know the techniques for carving one.

Water Supplies

I became a friend of the great geologist Chester Wentworth, who is an expert on water supplies as well as on all aspects of island geology. He was the principal scientist for the Water Department of the City and County of Honolulu and had studied a wide variety of geologic water resources and geomorphic problems in Hawai'i and published so much that he was well known. He invited several of us to view with him some of the water facilities supplying the island with water.

We started the tour at one of the main tunnels that were dug to skim fresh water off the top of the freshwater lens near Honolulu.

The surface of the island is somewhat above sea level. Over time, rainfall infiltrates the surface and tends to form a water table, but the zone beneath the island is permeated with salt water. As infiltrated water collects at the incipient water table, it floats on top of the salt. The salt water is slightly heavier than fresh water, a density of 1.025 compared with 1.000 for frest water. For the water table to thicken, the fresh water must displace the underlying salt. To have a water table 1 foot above sea level, it must displace 40 feet of salt water for $1/.025 = 40$ or $40 \times .025 = 1$. The outline of the fresh water deposit resembles a lens known in the literature as the *Ghyben-Herzberg lens*.

Wentworth led us into a tunnel that was at the foot of a mountain and close to sea level. We entered a long straight hall on one side of which was a shallow concrete trough constructed so that the crest of its wall was slightly below the level of the groundwater table. Thus, the topmost layer of the groundwater trickled over the wall into a channel of sufficient slope to carry the water along the length of the tunnel. With this small amount of water spilling over the concrete wall collected over the length of a long tunnel amounted to a considerable flow.

When the water is led out of the tunnel, it is pumped up to a storage tank from which it can be distributed by gravity to the many subscribers. There are several such tunnels on each island where they are locally called skimming wells for the obvious reason that fresh water is skimmed off the top of the water table.

Some are called *Maui Type* wells because when a tunnel cannot begin close to sea level, an inclined shaft leads down to the level of the water table where the skimming is done nearly like the case I described. The situation is common on Maui. It is clear that if a significant amount of water is drawn off and the water table falls, the lower boundary of the fresh water lens rises and salt water takes the place of the fresh.

The principles discussed here of the interaction of fresh and seawater apply to coastal as well as island settings. There is growing consternation about salt water intrusion in coastal areas and this information about the near impossibility to restore the salted water bodies is highly applicable. Probably the unique water-supply techniques in island situations make the local people more cognizant and therefore careful than persons on coasts where there are alternative sources of fresh water so skimming is not necessary. But as in many other aspects of resource development, overuse occurs through thoughtless or careless development.

For the area where the groundwater lens is utilized by skimming, if the withdrawal is greater than the replenishment, the zone of freshwater is thinned from the equilibrium a maximum of 60 feet and salt water takes its place. But if continuous replenishment of the freshwater gradually presses the salt water down, the rinsing effect would gradually clean the host rocks of salt. Wentworth computed how long it would take for the salt to be rinsed out by continually allowing more fresh water to infiltrate than is used. The answer was hundreds of years would be required. There is no mitigation that will replace the original condition. The knowledge of water occurrence on islands and along some shorelines is based on the concept of the Ghyben–Herzberg lens and was mostly developed in Hawai'i much under the direction of Dr. Wentworth.

In the water field the long-term consequences of overuse are not being sufficiently considered. Even more disturbing, there is an intentional imposition on the resource with the knowledge that the consequences will fall on later users, not on the immediate actors.

Unfortunately, there are many places in the world where the underground supply accumulated during the ice ages and is not being replenished. Use of this water is mining a resource that can and will be used up over time. I have argued that such a resource should be kept intact to be used only and as long as there is a direct and immediate crisis. Thus, it would be like a savings account in the bank.

It was at this time in 1948 that I was notified that my father had died in Wisconsin. I was not able to get back to the States to attend. It was soon determined by all the family that it was up to me to get Aldo's book published for I had already made many arrangements for it. There was work to do readying the manuscript for publication with the least possible editing or changing of the original wording. Finally, it was done and my good secretary, Jean Webb, typed it in final form. Interestingly, she got so spiritually involved with Aldo's words that later she personally started a small conservation project in her home state of Ohio.

Dad had been notified by Oxford Press of the acceptance of the final manuscript the weekend that he died. At least he knew that his masterpiece was to be published.

The Rainfall of East Maui and Research in Climatology

When Hawai'i was a Territory striving to become a state, there existed a relationship in the field of water that I have never seen elsewhere. The highly forested parts of most of the islands are too wet for agriculture and therefore had never been appropriated by private interests. It was owned by the Territory, in other words by the federal government.

The east mountain of Maui is composed of the now inactive volcano of Halelakela, and the northeast face that is exposed to the trade winds is one of the wettest places on earth. The central isthmus is low and fertile but lacks water so it was logical to lead water from the wet mountain to the lowlands where it could be used for agriculture. To do so was a feat of engineering that will be described shortly, and it was carried out by the companies engaged in agriculture. The unusual part of the arrangement was that the companies paid the Territory for the water. Contrast that with the large subsidy given to the irrigators in the western states after the Reclamation Law was enacted in 1905, immense subsidies that continue today. Yet the companies in Maui were still paying the federal government for water up to the time that sugar and pineapple culture was no longer profitable and was given up in Hawai'i in the last half of the 20th century.

For irrigation a collection system was developed through the construction of canals and ditches leading from the heavy rainfall slopes down to the flatlands. Because the lee sides of the mountains and the low-level plains received only a modest rainfall, the gradient from very wet to very dry areas is exceptionally steep.

The canal system was designed to intercept the many small to medium, natural channels. A long tunnel, interrupted only where it encountered a rill or valley, was dug around the flank of the mountain from near Hana to central Maui. It was drilled through the successive spurs and emerged each time it reached a tributary valley. Where it crossed the tributary a diversion structure was built to pick up water but eject most of the rocks that the water carried.

As water emerged from the mouth of the incoming flow it entered an underground square tube and was discharged having received an increment of flow from a tributary. The tributary water flows over an iron grate that screens off larger rocks as they fall through the grate to a pipe and discharges through another pipe to add to the main flow. Meanwhile the smaller rocks and excess water are separated in a stilling well to exit through a gate to the main tributary stream. The internal gate and the exit gate are controlled by a float gage that varies the exit opening depending on the amount of water in the continuous tunnel. The lower the water level the more water is permitted to flow into the tunnel. These extensive irrigation works, canal, and tunnels were built by men, most of whom were imported from China. So, there was a large population of people from Asia.

For the technical support of this intensive agriculture, the leaders supported the Hawaiian Sugar Planters Experiment Station that employed the best talent, well-educated scientists and engineers specializing not merely in agricultural science but in entomology, genetics, plant physiology, mechanical engineering, and other specialties. The community leaders that supported the Experiment Station were concentrated in five organizations locally called *Factors*, engaged in agriculture, shipping, factories, importing, and monetary pursuits. They were very conscious of new developments in handling crops, water, factory management, and application of science to these enterprises. Somewhat later than the sugar business, some of them developed the cultivation of pineapple, a crop that was not irrigated, but dependent on rainfall.

This crop also required the application of the highest level of scientific talent available for which they formed a research center, the Pineapple Research Institute of Hawai'i. Nearly all the Factors were engaged in the pineapple industry as well as sugar. The cultivation of pineapple also required insect control, physiology, specialized equipment for harvesting as well as the factories for processing and canning the fruit.

Sugar cane was grown on the islands of Hawai'i, Maui, Oahu, and Kauai. Pineapple was concentrated on Lanai, Maui, and Oahu. The most extensive canal systems for water harvesting were those on Maui and Oahu.

As soon as I came to the islands, I could see that knowledge of local climatology would be first priority. The local staff of the US Weather Bureau was small and rather staid and I found they were not in a position to help me very much. For example, they kept records of only a few rain gages and the forecasting was quite useless. I expected that the plantations had rain gages that were not reported so I began an inquiry of plantation personnel. There was indeed an extensive network and many stations had long records of observations. Because I was hired by both experiment stations, I had access to both sugar and pineapple plantations so I began a systematic search for the records and the locations on all the islands. The cooperation was excellent and with two helpers, we had an extensive inventory with maps, descriptions, and pertinent data. The published data included 827 gages.

The central parts of most of the islands except Lanai were owned by the Territory, and not in private hands. In the case of Maui, the ditches and canals originated on lands of the Territory and led the water to private lands on the isthmus, where large sugar plantations were located. The amount of money the plantations paid the Territory for the water had been determined many years ago and was based on the available map depicting the annual rainfall. The East Maui Irrigation Company with headquarters at Paia was a major purveyor of water for irrigation of cane and was the debtor paying the Territory. They suspected that the basis for the charge was incorrect and felt that the contract should be renegotiated on the basis of an up-to-date map of the annual rainfall. The Territory, represented by the US Geological Survey, and the companies agreed that a new map should be prepared by an impartial person, and because I was in a research station, they asked me to do the work, a proposal I was glad to accept as a part of my regular duties.

The request came at an opportune time for I had just completed my detailed key to rain gages in the Territory, so I knew which gages were available on East Maui. The rain forest on the flanks of the mountain of Haleakala, includes many deep valleys cut into its face, some having a thousand or more feet of vertical wall. The clouds cover the central face from ocean level to a few hundreds of feet at the cloud base to the cloud top controlled by a temperature inversion that usually is at about 7,000 feet elevation. The inversion is a layer of air a few hundreds of feet thick through which the temperature increases with altitude instead of the usual continual decrease. Above that there are essentially no clouds and so the upper part of the mountain, reaching more than 10,000 feet, is nearly always dry. Only when big storms cover the whole island chain does any rain fall on the mountaintop. The maximum rainfall occurs at about 3,000 feet. The inversion prevents the clouds from towering, and thus makes the top of the cloud layer essentially flat, a phenomenon often seen from a high-flying airplane. This inversion decreases in height from Hawai'i east to the California coast where the clouds near the seashore burn off in midmorning as the land heats up, a typical summer day in coastal California.

For use in constructing a map of mean annual rainfall, my map of gages showed there were on East Maui 116 rainfall gages in an area of 400 square miles, an average of one gage per 4 square miles, probably the most dense network in the world. Of these there were 13 with records of 30 years or more. Of course, there were fewer in the mountainous area than in the cultivated isthmus of the island. It was necessary to adjust the records of various lengths because failure to do so can distort the isohyetal pattern. To pick a long-term base record for station adjustment, both relatively windward and leeward locations were chosen. Five stations, each having records of 42 identical years were chosen, located with nearly uniform distribution geographically around the high rainfall area. The records of those five were used as the control and each other station had its record adjusted to approximate what its amount would have been in the same period as measured at the base five.

These adjusted values were used to construct the isohyetal map. But the key to such construction was the concept that the mean rainfall varied uniformly without steep breaks so a profile of values should be smoothly varying.

We had studied this matter and found that the mean rainfall varied with increasing elevation in a uniform way, in fact as the logarithm of distance. In the profiles on the slopes of East Maui, there was one anomalous gage with a rainfall value that did not plot the way one would expect it should. The gage called *Honomanu Mauka* with an adjusted value of 261 inches was considerably too low and it was close to the zone of maximum rainfall. There was no way of determining the cause of the anomaly without an examination of the gage itself and its location.

The trail that led to the gage was many miles long and through the heaviest rainfall zone. It was an opportunity of a lifetime. The manager of the irrigation company, Mr. Bruce, offered to help me see the actual situation on the ground because I felt there was something peculiar that might be ascertained by seeing it. He had never been in the center of the heaviest rainfall area and apparently wanted to be part of the examination. We saddled the horses at his headquarters at Paia and started up a long ever-rising trail that began in a grassland nearly without trees and then through a zone with lots of planted eucalyptus, into a mixed but thin forest with some koa that looked ancient and certainly was not reproducing. I had heard that in earlier years this slope had been very heavily grazed. These partly open areas were dominated by an exotic grass of the *Panicum* family that grows profusely wherever there is available sunlight, spreading by prolific suckering of roots. This quickly gave way to a real forest that became more and more dense as we proceeded up the mountain and it began to rain.

In the little valleys or notches in the slope the beautiful light green of the kukui trees dominate the scene, but in the main forest there is a combination of ohia lahua, lauhala, guava, some koa, and kukui. In the most dense part the giant native ferns are common. Through the constant rain we rode over the ground cover of mosses, shrubs, and small ohia lehua trees, the boles of which are covered with moss and the small native orchids.

Beginning where the annual rainfall is 10 inches, we climbed to an elevation of about 3000 feet and 400 inches. This was an experience very few people had ever had, even those who lived a lifetime in the Islands. Though the rain was constant, it was never heavy, as I had imagined it would be, and owing to the moisture we perspired a lot from lack of evaporation but we were never cold. There was a feeling of exhilaration that comes from an experience never to be repeated.

I found the gage and saw that it was placed about 50 feet from the lip of a pali, a cliff at least a thousand feet of vertical wall down to a distant stream that fell into the surf of the ocean. The location reminded me of experiments by my friend, the geologist Chester Wentworth, who showed that one rain gage at his roof-tree collected less rain than others normally exposed. The rain tends to blow over the ridge and give abnormally large values on the ground some distance to leeward. This clearly was the case here, for the updraft rising above the cliff blew rain over the gage rather than in it.

The rainfall maps drawn previously by the Territorial Planning Board and one by Stearns and McDonald had been constructed from far fewer data than were available to me. These maps showed two separate areas of maximum rainfall of 350 inches. My much more comprehensive analysis concluded that there was a single area of highest rainfall of more than 400 inches. This made East Maui among the three rainiest locations known on earth, more than one inch a day every day in the year.

Two decades later the commercial atmosphere has changed. Tourism had become the principal business. Competition from the Philippines and Brazil had closed down the sugar and pineapple industries of Hawai'i. With the death of the plantation economy the most notable rain gage network in the world was eliminated. The sprouting of high-rise buildings in Honolulu cut off the natural trade wind and the sea breeze that had kept the local weather always comfortable. The imposition of air conditioning is the greatest shame of a beautiful island.

But nothing can erase the deep impression I still retain from a simple journey into an area unknown and available only to a very few in the future. The aesthetic and science merged in a manner that lifts the spirit.

4

Travels Toward Science

While in Hawai'i, the science of climatology kept me both interested and busy but there were other things tugging for attention. I had the good fortune to have as a consultant on my island problems the greatest name in meteorology of that time, Dr. Carl Gustaf Rossby, professor at Chicago. In my long interaction with this genius I could see that my mathematics were insufficient to ever expect to make a name in the science of meteorology. Climatology was going to be less important as a science than meteorology and, indeed, during my career only a very few climatologists were elected to the National Academy of Sciences.

While I worked in Hawai'i, in addition to the papers I was writing on Hawaiian conditions, I was also writing up some of my observations of the past decade in Arizona and New Mexico. In 1949 I passed one of these manuscripts to my former professor at Harvard, Dr. Kirk Bryan, who had seen some of my published work so he knew that I was really was involved with science. In reply to my request for a review, he wrote me a one-sentence letter to the effect that if I came back to Harvard I could get the PhD degree that I clearly needed. Dr. Auchter had died and the interim administration was less impressive so it was without hesitation that I immediately started to plan to return to the mainland. I began correspondence with my previous associates in the Geological Survey. On my departure to Hawai'i in 1946, the Chief Hydraulic Engineer told me that if in the future I wanted to work with them, it would be arranged.

I left Hawai'i in summer having accepted a position with the Geological Survey, the officers of which kindly agreed to allow me to take leave without pay for schooling. During the time I was in Cambridge and in the intervening months I was assigned to the Survey office in Los Angeles. The Water Resources Division never had a position devoted to research prior to my hire so the local office in California did not know quite how to handle me. Thus, I was free to make my own plan of work. I took up a problem that had been started while I was in Hawai'i, the study of rainfall frequency in semiarid areas.

I also arranged travel to New Mexico where I was to meet some old comrades for a field reconnaissance of rivers that were typical of the semiarid washes in which water flow was episodic. I arrived by train in Gallup but my friends had not yet appeared so I put on my field clothes and walked to the dry riverbed of the Rio Puerco del Oeste to look at the geology and any terraces that might be there. Within a few miles I had mapped the terraces and the distinct formations in the exposed arroyo walls and now it was important to see if these could be identified in other channels in the vicinity. Within a few days I inspected several and it was in one of the first that I found a decorated potshard in the gully wall. This offered the possibility of an approximate date for the deposit.

When it was identified the single shard did provide a conditional date for one of the main periods of valley deposition in northern New Mexico. It was, of course, followed by multiple sites later explored, not just for their archaeological importance but for their geomorphological context. It also was just one of the times in my career that an unexpected chance encounter was to play a prominent role in my research. The resultant chapter in my thesis propelled an intensive interest in river terraces that has been a consuming concern throughout my life. I have no doubt that other scientists have had some similar experience.

The time at Harvard was very rewarding despite the shortness of my stay. In the five months I wrote my thesis, passed the orals and two language exams and took a class under Prof. Bryan. One of the most important things that happened at Harvard was the beginning of an enduring friendship with John P. Miller, also a student of Prof. Bryan who was to graduate at the same time as I. John was already an accomplished geologist who had mapped geologically three quadrangles of the Sangre de Cristo mountains. He was also a growing expert in weathering and soil formation. We decided to spend the field season after our graduation looking at the terraces and soils in eastern Wyoming.

There is a temptation to talk about the road to science or the paths leading to a career in science, but in fact, there is nothing but an unexplored landscape with neither roads nor trails. One thing that seems completely true is that the trail one takes should be a pleasure, that is, that each day one can look forward to going to work. There is today for many people little pleasure in day-to-day work, a situation that can be forged by many circumstances resulting in a lack of a feeling of accomplishment. Persons in science work hard but the long periods of detailed, repetitious work or

John P. Miller.
Courtesy of Mrs. John P. Miller (photographer unknown).

discomfort are overlooked, gladly accepted, and considered just part of the job. That seems to be a nearly universal feeling.

Of course, the enjoyment can be constrained if there are oppressive rules, inconsiderate superiors, or limitations on freedom. These are more discouraging than low pay, insufficient equipment, or other sources of frustration. In the field of water, as in many disciplines, learning begins with just looking, observing conditions in the field, noting common characteristics, measuring various parameters, and stating hypotheses that might be tested with data. But it usually begins by just living with it.

When one looks at the field trips that I have made and the obvious good times we had in work and in camp, it might appear that this was all a vacation and such trips had little to do with water or even with science. It may have been indirect but there was a lot of learning during these excursions.

Terraces of the Powder and the Bighorn Range

At a roadcut through the bluff overlooking the Powder River near Arvada, Wyoming, John Miller and I stepped out of the car to examine the exposed face of sandy silt that the highway construction had revealed. On the face was a network of narrow brown lines tracing the turnover of the surface soil exposing churning, "folding," compressing, affecting the top meter of the surface material. I think John and I simultaneously said aloud, "periglacial frost action." One reason we were surprised is that Arvada is many miles from the front of the Big Horn Mountains that were once covered with glacial ice. We immediately began a map of the location and a drawing of the figures we saw exposed in the cut.

That was just the beginning. To the south of this site and upstream we saw an exposure of a white zone about a meter thick well below the present surface, interpreted by us as a paleosoil, or ancient soil formed in a past period of aridity, a climate promoting the accumulation of calcium carbonate in the B horizon. So this landscape had at times in the past been both cold and moist when the mountains were glaciated, and another period warm and dry.

These findings led us to wonder what we might find in the mountain valleys from where the Powder River and its tributaries emerged. This river valley was bordered by three terraces, remnants of periods when deposits of sediment raised the elevation of the river bed, then followed by erosion of these same deposits, now only flat-topped spurs jutting out from the valley wall. The successive terraces were the results of climatic changes, alternating between periods of deposition and periods of erosion.

Recently, Professor Kirk Bryan, under whom we did our graduate work, had postulated that tracing such terraces upstream into the mountains might provide the basis for a chronology of climatic change in the Holocene, recent geologic time. The hypothesis was a logical one. If a terrace can be traced through a moraine and upstream of it, then the terrace must be younger than the moraine and younger than the glacial advance that deposited the moraine.

There follows from this a logical but untested assumption that two adjoining valleys ought to have approximately similar climatic history during the glacial periods. John and I decided to test this. The year before in Wyoming John and I had explored Clear Creek in the Bighorn Range all the way to Florence Lake at the foot of the cirque that is the head of the valley. We decided to make a similar examination of the nearby North Fork Piney Creek.

In 1952 the purpose seemed clear, the prospect interesting, and the experience both useful and enjoyable. The Wyoming mountains of the Bighorn looked blue and inviting from town, where we were anxious to get under way. Noon had passed before John and I finally pulled out, drove to the Hunter Ranger Station, and chatted with our friend from last year, Ranger Shulz, a tall, good-looking man wearing the kind of Levis that fit just tight enough, heeled boots, and a Stetson with a cultivated curl of the brim. Told him where we might be, borrowed a set of aerial photos, and took off. Right near the ranger station we ate lunch and sorted the outfits into our backpack. This is always a pleasurable job filled with anticipation. I looked longingly at a couple of oranges but John said that extra pound was verboten. The weight of the two packs was carefully divided equally.

When all was stowed, we drove up the road over Hunter's Mesa to the dude ranch inn, a place called Paradise Valley, a dream of a location. It is nestled in a wide part of the valley enclosed on all sides by timbered hills, the ponies grazed on a vale of green grass beneath the blue sky of the mountain country.

I went to the corral to ask the foreman where we might park the car so it would not be in the way. He and the dudes were equally surprised, disdainful, and yet a little bit wistful as they watched us throw the packs on our backs. His comment was something like, "I'd drop dead before I got out of the yard with that thing on my back. Give me a horse, any day." They have a pretty stream at the ranch, French Creek, but it must be about fished out we figured.

Across the creek and onto the hill slope in the timber was a trail for a few hundred yards but it petered out, so we cut into the timber far enough from the stream to avoid, if possible, the green jungle of the riparian along the bank. The country went up fast for we were on the steep downstream slope of a massive moraine. In the creek the water was just one rushing riffle after another, blocked with fallen timber.

Working our way up the forested mountainside was quite a chore because there was no trail, yet it was somehow satisfying. French Creek heads up to a narrow divide on a flat summit. We were on the crest of the moraine in what we called Soldier Park Stage in the glacial sequence. This moraine stretches uphill for miles and can be seen from afar by its steepness, standing above the undulating ridges and valleys of the lower country. Part of the North Fork of Clear Creek is diverted across the divide and dumped into French Creek, where it provides the major part of the flow. For this reason, French Creek seems to be flowing with more water than its channel was designed to carry.

Toward late afternoon we topped out on the divide between Clear and French Creeks, and at that place French Creek appeared to be a sort of canal, for embankments have been thrown up on the downhill side. We picked out a spot for camp in a copse of pines where we would look down and out over the lower country, but we could not yet see the mountain peaks ahead. There in Soldier Park we found a low terrace along the stream indicating the relative youth in geologic terms of this massive moraine.

While John rustled camp, I put up my fly rod and tried the trout but there was nothing doing despite the fact that the water was clear and cold. Darkness came over a snug camp and a couple of happy, tired mountain men.

I think I will always remember that first morning of this trip, pouring hot, strong coffee for John, neither of us talking, just looking and listening to the creek and some far-off Clark's Nutcracker.

After breakfast we left camp as it was, but put the packs well up on a pine out of reach of a bear, we hoped. Cutting north and east along the slope of the big moraine, we saw many ponds, kettles, boulders as big as houses, everywhere glacial features in good preservation. John told me many things about glacial landforms that I did not know, for he was experienced having mapped several quadrangles in the Sangre de Cristo Range. Also, he was very knowledgeable about soils, soil processes, and the interpretation of soils in terms of climate and chemical processes. It is no wonder that I found being with John in the field so interesting, entertaining, and instructive.

We went down the moraine slowly because there were elk tracks and signs everywhere. Suddenly, a hell of a commotion in the big trees ahead, as the elk herd plunged pell mell through the forest; we could not see them but they sounded like an army of elephants. No deer ever made a racket like that.

In a draw we found a rock river or boulder stream, an unusual phenomenon. On a very gentle slope, there was a ribbon of large rounded rocks, 1 to 3 feet in diameter. The rock river was about 40 feet wide and winding down the valley for more than a mile, meandering like a regular river. The rocks move down the slope by the action of freeze and thaw. We could hear the gentle trickle of water underneath the rock stream, that would be frozen in the winter and contribute to the slow and gradual advance of the whole stream of boulders.

To get a better look at this unusual feature, we climbed a pinnacle that stood above the surrounding big trees. We could see over and down into the park where Paradise Ranch lay. While on this high perch, we not only were fascinated by the visual length of the rock river but there were a number of rose-breasted cross bills perched on top of the tall spruce trees. We never could have seen them from the ground.

Returned to camp about noon; John fell in the creek crossing on a poor log but luckily was not hurt. It was a fast, rocky reach. Pushing ahead following the diversion canal, there was a swamp that lies in a flat place in the moraine. Two deer in the swamp looked at us briefly from 30 yards, and dashed off with a fawn following some distance behind.

The distance along the canal was much longer than we had anticipated but we finally came to the diversion structure thrown across Clear Creek to dump water into French Creek. Near the structure was a crude cabin for the water-master when he comes here. Up Clear Creek we shortly broke out into a broad, open, grassy flat, an old lake bed. Beyond it, the snowcapped peaks in blue and hazy splendor hung in the west as a great backdrop. It was all we needed. A big pine stood by itself on the edge of the park, and 100 yards away the stream came out of the willows and began the descent of the moraine front. We hung our packs on the great pine and luxuriated in the happy chore of camp making.

I caught a couple of trout, one of which was a golden that I had never seen before. We determined to improve our technique of cooking fish, making cooking time longer. They improved as a result.

Perhaps because the campsite was so pretty we decided we had business in this place. Using the full afternoon John went up the park while I fished. As I came back to camp, supposing John had already arrived, I left the buggy willow thicket near the stream and cut up to the broad flat of the park and spotted John standing in the middle of the big grass flat. He motioned to me to come but it was clear he had something of interest. It was a badger that now was just between the two of us. Caught out in the open in a short grass flat, he decided to stand his ground instead of running. We closed in to about 10 yards, where he was crouching with an arched back, hissing and showing his teeth. He worked his way backward slowly but defiantly but refused to run. Slowly John worked around him in a half circle until we stood together, and the badger, seeing he was not really surrounded, backed more rapidly until at 20 yards he turned tail and made it for the timber 200 yards away, keeping low to the ground but going like hell.

In the morning it was a push to Triangle Park and the Piney, then on to the north where the trail, where it existed, gets steep and we began a long, tedious uphill pull, working around the lower part of the Ant Hill that looms rocky and bare above us. The Ant Hill is a great dome of rock reaching well above 10,000 feet and dominating the local landscape. Hours wore on and the trail became harder and steeper. I was tired. We emerged out of the timber to a broad saddle that stretches east of the Ant Hill. When we reached the crest we ate chocolate and nuts while we looked down over the world.

From the crest the nearly obscure trail led down to Elk Lake, the border of which was too marshy to traverse. To get past it, we chose to walk, jump, stretch, or leap from rock to rock on a steep boulder field, and when we finally approached the lake we sat on a slight ridge to study the maps. A pack train passed us carrying a flock of uncomfortable dudes who looked both. All they wished for was that they could get off those saddles. Later a pack train came by carrying their duffel, including a new and expensive suitcase strapped onto a pack saddle. They must have had equipment for a regiment lasting a year.

Elk Lake is in a saucer-like flat on top of a moraine laid against the side of Ant Hill. We cut around the lake, left the trail, and headed NNW hoping it would get us to the so-called reservoir. It was rough going over the giant frost-riven boulders on a steep slope. The lake is long, about 2 miles, held up by a huge moraine. For regulation, a water company had put a fill of some 10 feet on top of the moraine and a water gate. The increase in water level killed all the lakeside trees so they stood as gaunt black poles.

Putting down our packs to look around at South Piney Creek Valley, we spotted a log cabin at the far end of the lake that appeared deserted. From the diversion structure the water tumbles down 150 feet to the toe of the moraine in a series of rapids and falls through the boulders. On the far side of this torrent, I saw a person who, through the binoculars, was cleaning fish. And fish they were: at least five 14 to 18 inches long. He had a long full beard, a floppy black hat, and dirty Levis. We sauntered over and admired the trout.

"Where did you find those?" I asked.

"Oh, jes' anywhere."

"What kind of bait?"

"Jes' about anything."

I put up my rod. It wasn't so simple. I didn't get even a rise on anything I tried. I worked down to the foot of the moraine where the torrent entered a small lake, Flat Iron Lake. Tracks on the shore showed this is where he fished, but I couldn't get a thing.

Meanwhile, John shot the breeze with him and when the fellow saw we weren't the regular brand of tourist, he opened up. For bait he used a small green plug called a *silverfish*, about three-quarters-inch long. Small fish were rising all over Flat Iron, so I even dug some worms and tried them. In the meantime, I lost my precious flybox, one Aldo had made many years ago in Albuquerque and had used all his life. I hunted for a long time and never found it.

We moved around the lower end of Flat Iron and found a dream of a campsite above a big pool on the moraine. I was out of flies and had little to work with so I pulled some wool out of a pair of my socks, and John donated the fancy red silk lining of his hat.

I tied some flies; I used a white feather we had found on the trail, tied a coachman variety Flat Iron, and caught enough trout for supper.

I woke up early for we were starting for the glacier on Cloud Peak. On the way as we passed Whisker's cabin, John spotted a weasel, carrying off the entrails of Whisker's trout. The weasel apparently lives near the old boy's cabin, and is a sort of pet of his.

The day before we had asked Whiskers which side of the lake to go up. There is no trail.

"W'al," he drawled, "I've been up both and I don't give a damn for neither—whichever side yer on you'd wish t'hell y'were on t'other."

He couldn't have been more right. We went up the south side, and no sooner we got off the moraine the slope down to the lake became very steep, covered with spruce, and underlain by a frost-riven rubble of huge rocks, the size averaging five feet, and many as big as a house. It was climb, balance, climb—every step was over a crevice waiting to break your leg if you fell. The slope was so steep that to look up it brought the conviction that the whole rubble slope would surely begin to slide.

The going was so tough we began to work our way uphill, hoping for a better surface, but it only got worse. As I got tired, I lagged behind John who was ten years younger than I. The trees became smaller, we looked down a thousand feet to the reservoir, and now we could see beyond Flat Iron into Frying Pan, a lake about four miles downstream. Uphill, it was still more than a thousand feet to whatever lay over the crest. Up valley, the mountain walls closed in and grew ever steeper.

After some hours we thought we would meet a well-defined moraine, but instead it was a steep, glacier-polished step, or knickpoint. There we took some rest in a small patch of grass and blue columbine, ate some chocolate, but saw that we had hardly started. We had come only three miles and it was another six or seven to the last lake. Clambered around a small lake, greenish blue, bounded by steep sides and spruces growing near highly polished and striated patches of rock. Another lake and still another. On a steep knickpoint where ledges held a few grassy step-like berms there were many tracks too pointed for deer—mountain sheep. No deer would live in this godforsaken rockpile. We crept carefully upward until we could see over the crest into the next lake basin. We scoured the boulder-strewn precipices on either side, searching with the glass. I had the feeling of being watched, the tracks and sign were so fresh. Possibly we had flushed them but they were never seen. To have sheep close by meant you were miles from nowhere, and damn glad to be there.

This seemed to put a fresh set of feet on my extremities because the country was suddenly transformed. We saw bear signs, maybe a day old. The enclosing walls got closer and steeper—we could finally see the far end of the cirque, a vertical wall facing us, black, sheer, and ominous. On this wall were patches of snow somehow clinging to the face as patches or as long strings of white hanging in the crevices of the vertical. A rain had started over the distant Frying Pan, yet we could look down the valley of Piney Creek under the cloud base to see the string of lakes, named Diamond, Emerald, Sapphire, Turquoise. I called them the *jewel lakes*. Down even farther we could see Flat Iron Reservoir, and Frying Pan Lakes—all strung out like pearls on a string, each a different shade of green to blue to silver with increasing distance.

Thunder and lightning close behind us as we topped the last moraine and looked down valley. Above us and to the north was the cliff, the saddle above which was Cloud Peak Glacier, a white triangle hanging on a black curtain of rock. The moraine on which we stood seemed flat and low but it stood a hundred feet or so above the lake at the foot of the cliff, a lake at least as large as Frying Pan. Distances and height could not be estimated so big is the wild. The cliff was about $1/8$ mile from us and rose about 1500 feet to the tip of the glacier. There was no possibility of us reaching it for the hour was late and we were not equipped for such a climb, but we could still stand in awe of the whole spectacle around us.

With some regret but with a deep sense of satisfaction, we turned into the intermittent rain showers and worked our way back, jumping from one great boulder to another. There was a boulder rampart blocking the valley ahead and we explored the lower end and climbed up on it several hundred feet. Its form was not apparent when one stood on it, but the exceptionally steep portions indicated that this was a pile of rock riven from the cliffs above it by frost action, a conspicuous form that dominated the center of the valley. It was a pile of debris triangular in cross-section but strung out along the valley axis. Large snowbanks were everywhere and we recognized this form as an immense protalus rampart.

When in glacial time these nearly vertical walls were buttressed by snowbanks at the base of the cliff while rocks at the cliff top were riven by ice and frost action, prying loose rock from the cliff that rolled down the snowbank to the valley floor. This was happening on both sides of the valley. When the climate ameliorated and the snowbanks melted, the rubble in the center of the valley stood as a long deposit with steep sides separated from the cliffs that gave it birth. Kirk Bryan saw small deposits of this kind and named them protalus ramparts. But no one had described one on the immense scale we encountered, built not with fine rubble but of rock blocks as big as a house.

In our notes we tabulated the sequence of forms we had seen in adjacent valleys and found them to be very different deposits, despite their proximity, and we recorded them in the downvalley sequence. The two adjacent valleys were quite different.

Though they both were blocked by the massive moraine that we climbed soon after leaving the ranger station, from there on to the top of the range they differed in nearly all details. Clear Creek ended uphill with a very small protalus rampart and very prominent rock garlands hanging on the steep slopes near the crest, but there was no series of lakes as on Piney Creek formed by multiple small terminal moraines.

We convinced ourselves that at least in this range of mountains, the glacial sequence for the whole range could not be determined by the deposits in any one valley, for even between adjacent valleys, the sequence could be markedly different.

We proceeded down-valley along the top of what appeared to be a smooth surface that turned out to be boulder covered as difficult as we had seen earlier. We were on a ridge of bedrock covered with boulders and bordered by cliffs, and hundreds of feet above the talus slopes. If we tried to get below the cliff there was a possibility we could not re-ascend the one chimney we saw, so we decided to go on at the high elevation until a saddle was found from which we could look over the terrain for a better route. We reached one and decided the present difficult route, on boulders, was at least possible.

Finally, we reached an area where bedrock cropped out in narrow part by grass ledges bordered with mature spruces two or three feet high. Interestingly, seen from across the valley this zone had looked like a timbered slope, the dwarf size not apparent. The moss under the little trees gave us the idea we were present-day Gullivers. These miniature parks look down thousands of feet to lake after sapphire lake. There were fresh and still-steaming bear droppings under every rock and reentrant. Finally, we intersected the top of the lower adjacent crests but found them to be an immense field of the same boulders, despite how smooth they appear on an aerial photo.

In the last couple of hours we bumped down over the slope of a big lateral moraine on the left side of the reservoir, and I mean bumped down, sliding, grabbing small trees to slow the descent. At the lake we cut across a peninsula but veered too far to the left and found ourselves on a long rather steep river of boulders, when we could hear a stream gurgling unseen under the rocks. This stream wound down to the north edge of Flat Iron Lake where we hit the trail and were in camp at near dark. Bushed after fifteen hours but buoyed up in spirit from what the wild had done to us inside.

I went down to the lower moraine and came back in less than an hour with five good trout. As we finally sat down beside the fire to watch the trout cook, damn if John didn't pull out of his pack a small flask of real whiskey. I could have kissed him, for even an orange had been a forbidden fruit.

The lesson learned on this expedition was an important one. The glacial sequence in two adjacent valleys was markedly different. Apparently, the local variation of precipitation, aspect, heating, and cooling is sufficient to produce different glacial histories.

The New Fork and the Green River

Spurred on by the wide variety of physiographic and topographic conditions in which river channels and their deposits occur, I felt it was necessary to find a stream not far from home that could be watched through time to see what changes occur and perhaps learn why those changes occur. I found a stream near Rockville, Maryland, about half an hour from my home that might satisfy. It is called *Watts Branch*, a small tributary to the Potomac, and drains what was at that time agricultural land in crops. There was a reach of the creek in a pasture that was open, pleasant, and easy to access. The local landowner had no objection to my studying there and we began by establishing a series of observation points along the stream meandering through it. Having established some benchmarks by driving nails in nearby trees, we surveyed the cross-section of the channel in bend reaches and in straight.

At the end of the season when I came back to look at the channel all my wooden stakes carefully established to mark the ends of the surveyed section had been chipped off by the mowing machine. So we started anew and marked each end of the sections with a three-foot-long iron rod driven so deep that the top of the pin was just below the sod surface. The pins were located by tape distance from several benchmarks and from other cross-section end points. This turned out to be a good strategy for when we came back in the spring to remeasure, the distances measured from the benchmarks marked a point where with a shovel sliced under the sod, a sharp ring was emitted when the shovel hit an iron pin. This repetition of surveys through different seasons and successive years began a documented history of change in a particular stream, not duplicated anywhere. During the many years of observation, the flow was measured at different times, velocity distribution recorded, bed material analyzed, and longitudinal profile surveyed.

Although at that time it was still rather hazy in my mind about what was important to do, it did seem clear that it would be useful to measure the principal physical characteristics of many channels in different climate geologic and physiographic conditions. The next trip to the west was aimed to do that. My colleague, Reds Wolman, and John Miller were recruited to participate.

It began with a night in an alfalfa field near Lexington, Nebraska, hot and swarming with mosquitoes. The following day we stopped at a number of stream- measuring stations on the North Platte River and that made things look more interesting. The river is braided, has many channels around vegetated islands. Near Ogallala we swam in an irrigation ditch that was the high point of the day. Below a concrete drop structure or check dam the water was deep and turbulent, the temperature delightful. Except for that we were willing to write Nebraska off our itinerary. Camped in a cottonwood grove near Torrington and cooked dinner as a rainstorm passed us by.

The idea was for me to show Reds the channel at Lance Creek where John and I had previously surveyed, and meet John returning from his Alaska trip. In the Lance Creek Valley we set up camp, I cooked some sourdough pancakes for breakfast, and we reorganized the outfit. After looking at the terraces of Tick Bend, below the village, we headed toward Hat Creek to find some badlands when I wanted to measure rills cut in the unvegetated hillslopes. South of Harrison a thunderstorm threatened but passed us by. I thought we might get some measurements if we found a gully that flowed water.

We finally got to Casper but dark and lowering clouds over the airport were discouraging. John was to arrive at 6:30 but there was no plane scheduled for that time, either incoming or outgoing. We decided to come back at 9:00 but because John said he was coming, we said somehow he would arrive, even if by parachute. Sure enough, an unexpected plane arrived at 6:30 and John was on it.

John's schedule required us to cut this trip short. A quick council and we decided to head for the hills, even in the dark and the rain. Ate a cold meal on a muddy side road, backing out from which we got stuck. I helped push the car out of the mud but an hour later I noticed the sheath knife I had made myself was gone. I was heartsick but we returned to the muddy spot, turned the headlights on where we were stuck, and John found it in the sticky track.

On again to camp out of the rain in the Riverton area where it was not raining. No wood in this sage plain but we found some old surveyors stakes along the road and I cooked sourdough on the small fire. After shopping in Lander we headed north to Burris on Dry Creek. We had heard that there was a road to the mountain crest where an Army camp of mountain troops was receiving training. The road was steep, high centered, with lots of hairpin turns but we emerged into a timbered park at about 9,000-feet elevation. The Dinwoody glacier was both distant and probably swarmed with army, so we decided to drop the thousand feet into Dry Creek valley. We bushwhacked down through the timber without a trail.

Dry Creek here was flowing through a broad reach with a few gentle riffles. John scouted for a campsite while Reds and I put up our rods. In a few minutes we had five trout accompanied by appropriate yelling and shouting. There were trout everywhere, but a short distance downstream the character changed to a steep rushing series of small waterfalls. We studied the characteristics of the channel as well as we could without instruments noting the distance between rapids, the length of pools and rapids, the change in size of material on the streambed, and the relation of these parameters to channel width. I did not realize until much later this brief experience led John to make his detailed study of mountain streams in New Mexico.

Camp was made on a big outcrop of glacially polished rock where pines and spruce had taken hold. Gathered needles and cut spruce boughs to make the bed that was in a little hollow in the outcrop. Below camp we looked down the steep valley to a broad riffle in the boulders, but next to camp the whole stream poured down a series of chutes and falls in a feathering mass of white water.

Though the mosquitoes were thick where we began to fish, as darkness came with the cold they disappeared. Dinner in a happy camp. Soup and a tragito followed by fish, macaroni, and coffee. The stream pounded in our ears as the stars watched us sleep.

Daylight broke clear and cold. After breakfast of sourdough cakes, we packed up and pushed upstream through glades of big pines, great rock flats of polished granite, little underbrush except where tributary creeks came in through matted alder, aspen, and willow. Even without a trail it was both easy and delightful walking. The wind was with us when we came upon a huge male elk that was rubbing the velvet from his horns. We moved quietly to within a dozen yards and even when he saw us, he moved toward the stream unfrightened.

Later we came upon a gravel terrace some twenty feet above the stream where we found many chipped red flint flakes, so we had an artifact hunt. John found a broken arrowhead of good workmanship.

We continued on without a trail. As we came closer to Dry Creek Lake, there was one big outcrop of polished rock where the stream plunged through a gorge about fifteen feet wide, in a series of falls below which was a pool of wonderful proportions. On a protected side behind the outcrop that overhung the point as a cliff, the green water swished in a deep small eddy, edged on one side by the frothing bubbles under the fall. As we looked down on the green eddy the dark backs and occasional golden bellies of big trout moved from the shadows of the riffle. There was apparently no dissent as we put down the packs and unsheathed the rods.

Reds got onto a beauty and after playing him for twenty minutes amid much whooping and hollering, John snagged him and brought him in. Even John, no fisherman, got excited and attached a spinner on a willow stick, and proceeded to work the fast, deep chute above the fall. There was a big strike and the leader snapped. Finally, John borrowed my rod and landed a beauty. When the sun told us we'd better move, we had a string of fish and a lot of things to laugh about.

It was on this trip that I began to analyze some of my cooking procedures. I had learned from my father that good camping means good cooking and I worked at it. Hot food should never be served on cold plates and in all the meals I served in camp—rain or snow—this was a strict rule. Because I did all the cooking for the people in my camps, I had many recipes in my head that were tasty and nutritious. Breakfast always included sourdough pancakes and as long as I had a Dutch oven, they were always thin and delicate. Pancakes were generally served with bacon or sausage, and sometimes an egg.

We always had boiled coffee, strong but never bitter. When on a walking trip and cakes were made in frying pan, they were good but thicker. The starter I kept always and used in the winter but never went to the field without it. Where we had a camp for some days in a row, I made sourdough bread but that was not usual.

In this camp on Dry Creek we were catching some large trout and I was not used to big fish. I had learned that the smaller fish were better if cooked slowly and longer, but for these trout, 15 to 18 inches, I decided to fillet them. A sharp knife cut made along the top of the back just passing the dorsal fin allowed me then to cut the side along the length of the fish parallel to the backbone and lift off a filet with no bones attached.

The best fish I ever tasted were catfish I caught in the silty San Juan River in Utah, skinned and filleted as here described and cooked quickly on a very hot greased iron plate.

We continued on above the lake to another wide, lake-like reach where we made a discharge measurement, 90 cfs (cubic feet per second). Lunched there and proceeded upstream less leisurely. The available map was not sufficiently definitive to indicate where we were. It was after 3:00 p.m. when we broke out on still another lake from which we looked out into a large cirque to the NW. Indian Pass lay well south of the cirque and Chimney Rock WSW.

As is the usual case we reach the most distant point in midafternoon, but the downhill return is always about twice as fast. Got to camp in time to catch some fish for dinner. The most distinct difference between the upper Wind River Range and the Bighorns, where we have explored in other summers, is in the character of the big rock areas and mountaintops. The Bighorns are covered with rubble of large size—the Winds have no exceptional talus slopes, and in fact, talus is not conspicuous. Few large areas of polished bedrock, a distinctive feature of the Wind Rivers, occur in the Bighorns. Here, isolated or scattered rocks and boulders on polished and striated surfaces call attention to the widespread occurrence of clean, little-weathered bedrock. Weathering appeared generally similar, everything looked moderately fresh and nothing was deeply decomposed.

Walked out of the Dry Creek canyon on a long tiresome trek. Reached the car and after miles of a difficult high-centered road, had lunch. Stopped at the big canal, a diversion from Dry Creek, for a bath and general washday. It is a shame that this beautiful river is essentially destroyed and led away in concrete. Drove John to the airport at Riverton after which Reds and I found our way in the dark to my old campsite on the Middle Fork of the Popo Agie.

Continued west the next day, made discharge measurement on the Little Popo Agie at the gage, and lunched at South Pass.

South Pass and the Sandy are of particular interest because the tracks of the covered wagons bound for California or Oregon are so well preserved one might think that they are still being used. A little distance from the Pass and in sight of the Oregon Buttes is the Parting of the Ways, where the wagon trail divides into the route to Oregon and the route to California. At Buckskin Crossing the Oregon Trail is spectacular.

We met the famous geologist William Rubey at the Sargent's Inn of Daniel Junction. Bill was the author of many outstanding papers in geology, had been the president of the Geological Society of America, and recipient of the Penrose medal. He was at this time still mapping the Wyoming Range, a project to which he had devoted many years. We had a drink and a family-style dinner, talked rivers and early western history. The old hotel is now long gone. Next day we measured the New Fork at Jenkins' ranch, the Bar Cross, at that time still owned by the Barlow family. Mrs. Barlow was a Jenkins daughter.

By this time, I had decided that there is a need for a calculation of a few simple measurement data from as large a collection of rivers as possible. At each location we should measure discharge, channel slope, size distribution of rocks on the bed, surveyed cross-section, and sinuosity. Drainage area would be measured from a map. This turned out to be a useful and informative goal, for soon after our data collection was published, Canadian engineers engaged in a similar task and published an even longer and more detailed description of Canadian rivers. We were now engaged in contributing to this collection.

Our procedure continued with a measurement of the Green River at Warren bridge. Camped on the upper New Fork. A pine squirrel came up a log where I was sitting and looked me over from about three-feet distance. Two Clark's nutcrackers nearly took my hat off. Camp was on a gravel-and-sand terrace about twelve feet above the stream but below a still higher terrace another fifteen feet above us. These cut through a morainal topography. The nutcrackers are fun to watch and call out a rather grinding r-a-a-a-ch on a rising pitch, repeated every few seconds. Fringed gentians carpet the low bar where we wash.

After completing our measurements near camp, we proceeded to Horse Creek, which joins the Green near Daniel, but only after flowing parallel to the Green for many miles. Stopped at the site at old Fort Bonneville established in 1832. Bonneville was apparently trying to get established in the fur trade and compete with American Fur Company. There is still a question of whether Bonneville was actually a spy for the American government sent to counteract the activities of Walker, a Scotsman, possibly sent by the British, who were trying to establish a claim to that part of the continent. Horse Creek near Daniel is important because it was described by Peale in 1879 as a stream of two distinct characteristics, it parallels the master stream, the Green River, for some distance before joining. He also introduced the term *anastomosing* into the geologic literature to mean *braiding*. He sketched a reach of channel that we explored and mapped, and we showed that the same island he mapped exists now nearly unchanged. We interpreted this to mean that the braided pattern could be stable with relatively minor change with time.

We mapped another reach of Horse Creek that showed clear evidence that islands of a braided stream begin as a gravel bar deposited near the center of the stream and in time become stabilized by the growth of willows or other species of brushy plants. Late in the day made camp by a meander reach of Duck Creek near Cora and measured cross-sections in both bends out at the crossovers. We returned to camp as the mountains turned pink. I had a new Dutch oven that allowed me to do some fancy cooking. Drinks, then soup, round steak with onions and tomato, baked potatoes and coffee. Long discussion of morals and ethics. The night was cold and clear.

In the morning my little friend the pine squirrel took the first pancake that I laid out for him. The new Dutch oven is now employed to make pancakes for all three of us. Off to Cottonwood Creek, which flows out of the Wyoming Range into the Green. The streamflow gage there was a beat-up affair that resembles an outhouse. It is placed close to a long series of riffles and islands, but just upstream is a set of high-amplitude meanders. So here is a place a meandering channel becomes braided. We mapped it with enthusiasm, made pebble counts, and cross-sections. This took all morning and we lunched on the streambank amid a garden of gentians, each petal of which has a small point instead of the usual fringe. We inspected North Piney Creek but it suffers from too much diversion and the pattern is not distinct.

We started a plane-table map of the Green at Daniel Bridge where the distributaries really meander and found that the slope of a distributary is steeper than that of the undivided channel. Later we could see that this is logical and expected, but at the time it was new concept to me. In camp at dark. Very large number of shooting stars.

Luna B. Leopold at work with his plane table.

Morning made me wonder what antics Chippy would produce today. He didn't disappoint. During breakfast he came right up to the middle of things and made off with a large chunk of sourdough bread that I had put out for toasting. Later he got in the car and started to look through the chuck box by way of general inspection.

We continued plane-table mapping on the Green. Though the mosquitoes were scarce, there were hosts of horse flies $3/4$-inch long. Were it not for some bug spray, we could not have continued mapping. We measured cross-sections of both channels and took float measurements for velocity. The sum of the channel discharges in the divided reach are not equal that of the combined flow. The trouble lay in the great depth of the main channels that we could not wade, so our sections are not sufficiently accurate. To correct this, we gaged all channels by current meter and recomputed. Came back to camp early, tied some trout flies, set sourdough for bread, and went fishing.

Next morning Chippy was late, presumably because it was cloudy. Later he came to collect the sourdough put out on the log for him. We sat in camp to compute, plot data, and summarize the field notes, Chippy in constant attendance. We broke camp and he called from the nearby pines.

Headed north to Hoback where I took some velocity measurements in the vertical to plot the flow distribution from surface to streambed. Measured velocity in the Snake River using floats in a rapids reach—7 and 13 feet per second. Later made a more standard measurement in a uniform pool reach, obtained the width by triangulation.

An unsuccessful hunt for a campsite on the Gros Ventre and ended up near dark in a good location on Pacific Creek. After a sourdough breakfast, went north to Togwatee Pass, where we surveyed a beautiful meander, then on to the Buffalo Fork. Lunch at the Pass in a little opening where lupine and paintbrush were mixed with other flowers in great profusion. A mother moose and newborn calf crossed the road. He was the most gangly little creature imaginable, with his thin legs that seemed to give out from under him with every step. The season was over.

A Plan for Research

One of my precepts is that part of every research program must be an activity that makes you look. This summer's fieldwork showed me that this was the right track. But it was equally clear that there were several fronts in water-related studies in geomorphology that must begin. I had had 15 years of experience with a combination of field observation and research. I felt qualified to lay out a program for myself.

First, it was necessary to study the morphology of the channel itself, the processes, parameters, the influence of erosion and deposition and their relation to hydraulics.

Second, there must begin an observational program in some local river channel that would permit direct observation of these same parameters, hopefully over a long period of time.

Third, we must have a compilation of relevant characteristics of a large number of channels in various geographic settings.

The first of these can begin with a perusal of the literature in engineering, hydraulics, and geology to find what is already known. The second must begin with choice of a nearby stream and an observational program for it. The third would take the form of visiting in the field as many different rivers as possible and a schedule of observation to be made at each.

The second of the tasks was already under way. As mentioned earlier, I had found a stream not far from home that could be watched through time to see what changes occur and perhaps learn why those changes occur. Watts Branch, a small stream near Rockville, Maryland, was chosen as a project that had no end in sight, for it would take some years for the stream to make the gradual changes that I was anxious to follow.

Our measurement of these rivers in Wyoming was the opening contribution to the third item.

With regard to the first of the tasks, I began a literature search for data on rivers and within a few weeks I discovered that in the room next to my desk, was the greatest collection of observational data on rivers in the world. It contained the summary sheets of the original data by the men who measure the flow of rivers throughout the United States. On these field sheets were entered the measured quantities of width, cross-sectional area, and mean velocity for every river measurement made in the past 40 some years. And apparently no one had heretofore used the collection for research purposes.

It was immediately apparent to me that these extremely valuable records did not include some of the parameters that would be needed to complete a description of the conditions at the time of measurement.. Not recorded were a measured cross-section of the channel though the necessary measurements had been made, but they were entered on other sheets not in this file. Nor did it include the slope of the channel or the water surface. Also, the material making up the bed and margin of the channel would be useful but were not observed. Even at this early stage of the program it was apparent that the delineation of the bank-full stage at each cross-section would be important even though at that time we had not even determined how it should be defined. These needed additions could be added by summer fieldwork during which we would go to stations one by one and make the measurements.

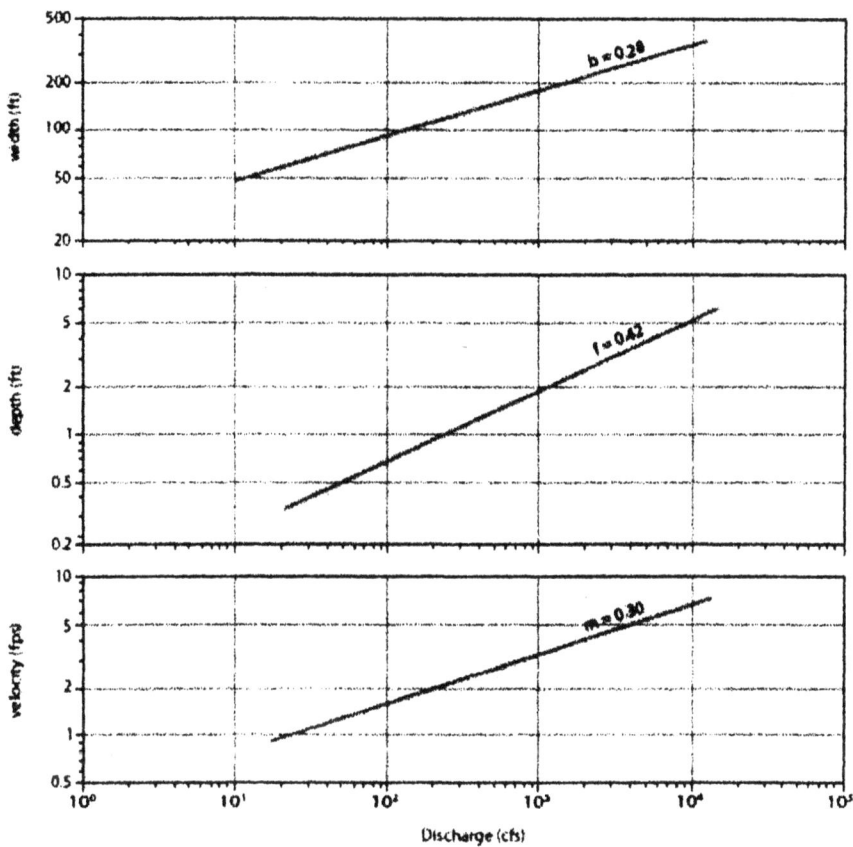

Typical relationships among river width, depth, velocity, and discharge at a gaging station.

Unfortunately, after the first years I had used this valuable collection, it was decided by someone to move the files to some place in the archives and even the employees of the Survey were not sure just where they were. This made getting the data a frustrating chore even when one knows exactly what he needs.

Hydraulic Geometry

Task number one seemed at first blush to be quite specific. To narrow the problem, I went through all the field data collected that first summer and began writing a report on the work. Though the emphasis was on the river terraces, the soils and probable history, my attention kept coming back to the channels we observed. How different were those cut in unconsolidated silt from those in the east where vegetation was a dominant feature. I decided the main question to ask concerned the width of a channel, what determined it, and how does it relate to other geomorphic parameters. It turned out to be a crucial decision. The question was very simple but it was one that would presumably be related to a host of other parameters.

Within a couple of months I had fleshed out the main features that relate the width of a channel to its depth, velocity, discharge, and drainage area. I described these relationships as series of power functions. I took the unfinished manuscript to my friends, John Hack and Charles Denny, who were impressed but they advised that unless the analysis included the sediment load, it would be considered interesting but useless. This added a whole new dimension to the problem and one in which I felt less than confident.

Because we had worked together on various sediment problems, my former supervisor, Thomas Maddock, was the logical man to consult. He was helpful and full of ideas so we decided to combine forces and work together on the manuscript. We brought it to conclusion after a few months. It was published by the Geological Survey in the Professional Paper Series, though up to that point no member of the Water Resources Division had ever been allowed to publish in that medium. The paper created a lot of interest, both among engineers and geologists. The geologists were particularly struck by the demonstrated fact that the velocity in a river did not decrease downstream as the slope flattened, but actually increased under many conditions.

It was in September 1964 that I had a telephone call from the office of Senator Clinton Anderson of New Mexico instructing me to be at his office at a certain time of the day. I arrived at the specified time and met in his outer office a group of men with whom I was well acquainted, David Brower, Howard Zahnheiser, Ed. Graham, and Bernard Frank. We were admitted to his private office and after greeting us, the Senator said, "Gentlemen, tomorrow I am going to introduce a wilderness bill. I want you to know, however, that I could not accomplish all the things we would all like it to have. Among other things, the bill will allow mining to continue in designated wilderness until 1982." I think there were some other minor things he mentioned but I do not recall them. It was a day long awaited, and one for which many people had spent years of work, especially Howard. Aldo Leopold would have been gratified.

Flood-Control Problems

The paper on the hydraulic geometry published in 1964 became well known and it led to some unexpected developments. Partly due, I presume, to the success of the paper, we were asked to write a book for the Conservation Foundation, analyzing the efficacy of various schemes for the control of floods, especially comparing the effect of a large dam as compared with a series of small dams. It took about a year but we produced an interesting analysis in a book entitled, *The Flood Control Controversy*, which contains a detailed analysis of the flood-control activities of the Corps of Engineers and of the Soil Conservation Service. In it we showed by hydraulic and geomorphic analysis that a dam of any size exerts an effect on flood discharge for an amazingly short distance downstream. Therefore, the large dams on the Missouri River, for example, have only negligible effect on floods of the Mississippi River. We showed further that small dams, even in large numbers, would not control large floods.

One result of the publication of that book was an invitation to go to India to see if we could offer any assistance to the government of India as they were constructing one of the largest levee systems in the world to control floods of the Kosi River. This assignment to India was an unusual experience for we were to see from the inside how a large and expensive project was developed and prosecuted. The following text describes some of what we learned.

India: Water Development Viewed from the Inside

We were traveling in an open jeep between Muzaffapur and Sitmarhi in northern India. As we approached a small village I saw a great crowd of people in the center and I asked the driver to slow down so we could see what was happening. As we approached the crowd they all turned toward our jeep and waved wildly, at the same time yelling "America! America!" I now saw what drew the whole village.

There was a young man pumping vigorously on the handle of a small red pump from which there was a modest gush of clean-looking water. A few days earlier, members of AID (Agency for International Development) had sunk a shallow well. These people had never before seen water that looked clean and they were filling all kinds of utensils while exulting in the wonder of it all.

One of the truly remarkable accomplishments of engineers in America and in Sweden is their provision of drinkable water for 100 percent of their citizens; several other countries have come close to that. Our country only recently realized that there is contamination in many of our sources that were unknown before the passage of the Clean Water Act, but these anomalies do not detract from the overall success of the program.

Travelers from the United States are at the mercy of what they find in the countryside of other continents. The large cities in Europe and the United Kingdom have trustworthy water supplies in hotels and pubs, but in many parts of Africa and Asia clean water is either a rarity or nonexistent. Many travelers have learned to live on tea, beer, or wine, but tea must be made with boiling water. The experienced ones require that they see the water boil for the tea.

Trying to learn some details of water development in India was a challenge and an experience for both Tom Maddock and me. The flood-control project is aimed to reduce the damage from the Kosi River, a very large drainage basin that begins both east and south of Mt. Everest, flows from the high mountains through the foothill zone, and debouches on the wide, flat floodplain of the Ganges. The Kosi River is only one of several large tributaries to the Ganges. It is an important cause of damage to the state of Bihar because it is braided into multiple channels and thus is very wide, and this river has moved laterally 75 miles in about 100 years by cutting its banks while depositing on the opposite side. With a population of 336 persons per square mile, the same density as Japan, the floods have taken a large toll on the economy and the people.

The Ganges is a truly impressive river, the eighth largest in the world and of the same order of size as the Mississippi. In this part of India, the river is so large that oceangoing junks look like toys. The river has many large stable islands, some of which are cultivated or wooded. In the low-flow season the emerged bars of white sand are separated by various channels.

The broad floodplain of the Ganges is frequently flooded, the perfectly natural way for a river to behave. What dams exist are so far upstream on the tributaries they have no effect on the main channel. As a result, all the small villages near the river are surrounded by an earthen berm or dike, which during flood makes a little island of houses and trees completely surrounded by water. Getting to see this immense country of which the western half is a desert and yet contains a dense population, gives a visitor from our well-watered country a new view of the value of water. Just getting to India is an education in hydrology and climatology.

As we landed in Karachi, we got a view of the dry Indus, the 23rd largest river in the world. Its origin in the western Himalayas provides it with ample water, but along the 1900 miles of its length it is progressively tapped, dammed, diverted, and degraded, until it reaches its many mouths with very little water remaining in the channel.

This part of Pakistan and the western part of India are spotted with a shrub like our mesquite (prosopis), bushy and thorny, as befitting this desertic land. The area is broken into odd-shaped plots of irrigated land and on higher ground, U-shaped areas of floodwater farming located on ephemeral washes. Many of these farms that depend on the infrequent flows appear ancient and are partly covered with drifting sand.

The airstrip near a village, if it exists, has a simple building where passengers may wait and get a cup of tea. These structures have thick walls and high ceilings needed in such a hot climate. The one we visited at Jodhpur was cooled by having the wall facing the prevailing wind made entirely of small sticks or brush over which a trickle of water flowed, and the breeze coming through the wall was cool and refreshing.

Outside the terminal, ragged turbaned natives were breaking up rock by hand, to be used in repairing a road. At the same time a Packard drove up and, dressed in a bright-red sari, a shapely woman stepped out and the driver unloaded expensive luggage.

Inside, workmen were plastering a wall. The plaster was mixed in small bowls that were carried on the head of a thin man who climbed an old, patched ladder, the top of which had been extended by tying a bamboo section on with thin cord. On the floor sat a ragged fellow bracing the ladder with a calloused foot.

An Engineering Office

We landed at Delhi in the middle of the night and it took a long time to explain that my two cameras were quite harmless. In the morning the work began. Here we were dealing with well-trained men working in an environment different from anything in our previous experience. Our task began with conference at the Central Water and Power Commission. The office is at Bikaner House, a palace built a few years ago by one of the local maharajas. The office of the chief engineer, Dr. Rao, is located on the second floor, entered only by a narrow flight of stairs leading from a small court in the rear. This was designed to keep the harem on the second floor separated from the main living quarters of the potentate downstairs, but still available. All the windows of the second floor were shuttered or latticed with extraordinarily beautiful stone screens. Each one of these works of art must have consumed a lifetime of an artisan.

The whole building was dirty, scarred, and unswept, and one wondered how these men worked in such a surrounding. Dr. Rao was a short, slight individual with dark, nearly chocolate skin. He was wearing a rather dirty, white cotton short-sleeved shirt and rumpled white cotton pants. He operated behind an extremely large desk covered with green felt. He was continually harassed by apparently unimportant things such as calls for a wrong number, as well as by important documents he must inspect. He spoke sharply to a host of turbaned underlings who did not seem ever to catch up with his requests. But he was very pleasant amid all the responsibilities and, in our short stay, he even planned our schedule to see various subordinates as well as the local sights.

One of the major deficiencies in the development and construction program in India and in many other countries is the lack of reliable data, quantitatively adequate, sufficiently distributed in space and time, to give information for efficient use of the water resource. Measurement of the flow of rivers of such large size as occur in India is difficult, time-consuming, and costly. A program in such a large country requires a network of observation stations, which means a widely dispersed corps of trained observers and a modern organization to check, summarize, and publish the data. The situation in that country fails to meet an adequate level. Because this is an important aspect of development, our work in India included learning about the practices, organization, and degree of dependability of data.

I asked Chief Engineer Rao if we could see the data-collection effort and his quick reply was not only affirmative, but he immediately used the little bell on his desk to call one of the several turbaned servants who was ordered to take us to the chief of the hydrology section. This position was comparable to the Chief Climatologist or Hydrologist in the counterpart organization in America. The Chief of the Section, Mr. Ahuja, was personable and talkative. He was educated in London but spoke with quite an accent. He was lucid and expressive as he showed us maps of the rain gage network and the stream flow stations, which seemed to indicate lots of measurement data. We found later that this was somewhat an exaggeration.

Mr. Ahuja sat behind a large desk slightly smaller than that of Dr. Rao, but also covered with green felt. The desk was nearly clean of papers but had a small bell. I soon found out that everyone in the hierarchy sat behind a desk covered with green felt, the size of which depends on where in the pecking order the person stood.

We were to come back to Mr. Ahuja, but now we spent some time with the engineer in charge of the Design Section, Mr. Kumra, with whom we discussed the general plan his group had proposed for the levee system. Mr. Kumra showed us a relief model of the project and many details of importance. Tom and I could see that the matter of sediment transport in the great river would present problems. In the foothills of the Himalayas where the geologic formations tend to be soft sediments, the valleys cut by the river have steep walls, the native timber has been seriously logged, and landslides are common. This combination produces an exceptional sediment load in the river.

Our most important assignment was to talk with the Chairman, Central Water and Power Commission, Shri Konwar Sain, whose post is comparable to a cabinet member of the administration. He was a very polished gentleman who spoke in a low but authoritative voice. The conversation soon got down to brass tacks; he specifically wanted our final report to indicate that zoning was important and that levees would be more efficacious than a program of dams. Our opinion was the same as his so we had no trouble in agreeing on these points. He also wanted to make certain that we omit anything that might cast aspersions against the Commission.

Poona Experiment Station

One aspect of the hydraulic design was handled by the famous hydraulic experiment station located south of Bombay at Poona. It is one of the largest and most active laboratories anywhere, known for the physical models of large size used to test various effects of different details in the design. It is the counterpart of the Mississippi River Commission Laboratory at Vicksburg, operated by the US Corps of Engineers. We flew to Bombay on the west coast, four hours by air from Delhi. It is a hot and humid climate and I saw all the vegetative types I knew in Hawai'i, banyan, mangoes, monkeypod, Plumaria, panex, croton, and papaya. We walked many blocks through the center of the city, stepping over dozens of homeless men lying on the sidewalks. We got as far as the Gate of India at the harbor, a Victorian arch that reminded one of the period when the British Empire was in full bloom. The center of Bombay is alive with shops, traffic, and a plethora of people.

There was a policeman in the center of a busy intersection that was a real show. A pink, well-folded turban topped a handsome face graced by a pair of mustachios that would do a Cossack proud. He wore a neat, pressed uniform of short sleeves and short pants above calf-length stockings that had bands of bright-red color. Most important was a leather Sam Brown belt with shoulder strap, on the belt of which was a small leather cup or socket into which stuck the shaft of a large umbrella, thus freeing both hands and arms to make sweeping gestures with expressive fingers, that kept the traffic of both cars and bullock carts moving through the intersection.

Wandering down toward the south end of the peninsula past beggars, bazaars, and people sleeping on the walk, we got to the King Edward Museum. The visitors who came to look were more interesting than the objects on display. Every kind of person came, from obviously wealthy to ragged and dirty, mostly in bare feet as is common. One Moslem girl was in a burka, head and body totally covered down to the waist on all sides like a straight sack, but with a cheesecloth over the eyes. She would face the cases, lift the veil to see, then draw the sheet down and move on to the next showcase. Everyone was respectful of the objects; the obviously poor were just as interested as anyone.

I was interested in the common mode of dressing. The Bombay state appears to be characteristically peopled by a dark-skinned, slight, and not-very tall race. The usual way for a man to wear the scarf-like gauze or sheet called the *longue*, is to wrap it around once, then draw it between the legs from front to back, and tuck the end in at the small of the back. The longue may be drawn over the head.

The sari is a long narrow band wrapped around the hips, then over the chest to one shoulder, thence around the back or over the head. The chest is covered with a short, tight vest or blouse that leaves bare an attractive midriff. The sari is tied in a knot in front after being wrapped around the hips.

The train from Bombay to Poona winds through the mountains of the Ghats. The cars have cubicles like those on British trains, but outside the cars the whole train seems to be crowded with people hanging on in every place where there is a handhold. These people seem to be just hitching a ride.

Whereas Bombay is built on a peninsula that juts south into the Persian Gulf and is hot and humid, Poona is near the top of a steep escarpment of the Western Ghats that falls down to the west coast. It is dry all year except for the three months of the monsoon when it receives only about 5 inches of rain. We were there in March one of the hottest months when the diurnal temperature goes from 80°F to 110. The city has a population of about 7–8 lakhs (1 lakh = 100,000). It has many neighborhood bazaars as elsewhere in India, all confined in a large metropolitan area.

The schools and university in Poona appeared drab and run-down, comparable to what we experienced in Delhi. Even beautiful houses seem to become decrepit with no paint or plaster, and with broken windows. The central building of the hydraulic laboratory was a good example.

We were certainly treated well for we were met at the rail station by the Director of the laboratory, Dr. Jugelkar, who drove us to the hotel for lunch. We spent a hot but fruitful afternoon in a tour of the experiment station. There were acres of flowing water guided by carefully modeled banks, duplicating on a small scale various river reaches being altered for construction of irrigation outlets, some constrained by levees, and others by bank revetments.

I had recently been deeply engaged in laboratory experiments using a 60-foot flume in which sand was moved by a very small flow of water. To feed sand into the water I had a container of dry sand, the bottom of which was equipped with a vibrator so a small trickle of sand constantly was fed into the stream.

Here at Poona similar experiments were being carried out on a large scale and the constant sand feed consisted of three men kneeling down near the head of the flume dropping sand from their hands into the flowing water. The contrast made me smile.

As we walked from one experiment to another, we were followed by a bearer with a tall pink turban, carrying a large umbrella, which he tried to keep over our heads to intercept the hot rays of the sun. Because we moved here and there, he had a hard time keeping the umbrella over us and he moved nervously from Tom to me back to Tom. Poor fellow, he just couldn't keep us both in the shade.

The main conclusion we reached from our tour of the big laboratory was that there was no attempt to quantitatively measure either input or output of either sand or water but depended on making slight changes in the placement and angle of various parts of the model to see what was the result. Thus they were led to a physical combination of details that have the desired result, and this combination was essentially duplicated in the field in an actual river or shoreline.

Water is an important resource in this large country, which has a long dry period following the annual monsoon rain. The principal source of water in the northern half of the subcontinent is the flow of rivers emanating from the Himalayas.

India has a long and distinguished history of constructing large-scale engineering projects for irrigation. The early work was done by Englishmen, but the native engineers learned much and they no longer depend on help from outsiders for ideas and instruction. The main works are barrages (low dams) constructed across a river, providing a head of water to flow down large canals that serve areas very distant from the source. The hydraulic design, construction, and maintenance of these great canals flowing through easily eroded formations have been finely tuned from long experience.

The canals that feed water diverted by low dams have served the country very well and the extensive experience is not duplicated in any other country. But in recent decades the government has felt the need for large and high dams mostly for the generation of hydropower. In this realm their experience is far from the level necessary for adequate protection of related resources. But India is not alone in this for countries all over the world are planning and building massive works that have a multitude of destructive effects. Not only is the choice of building high dams questionable, but equally important is the fact that the World Bank and other fiscal agencies, supplied mostly by the United States, not only finance but encourage these destructive programs.

The adverse effects of large dams are so widespread that millions of people are seriously harmed. I enumerate a few of the destructive results. A most important one is the displacement of whole communities from their native locales to places distant and inferior in productivity.

Below a dam the natural fluctuations in flow, from low to high to low, are changed or eliminated, and a channel so altered becomes in disequilibrium. This leads to bank erosion, scour or deposition, and serious alteration of riparian borders. It also often causes the destruction of headgates or diversions downstream, the drowning of historical sites, monuments, and cultural treasures. An important effect is the change of habitat of various species, especially loss of fisheries on which populations depend. Dams cause sediment problems of various kinds, usually unforeseen.

Looking for Data

Returning to Delhi, we wanted to pursue the question of basic data and we were anxious to see some actual tabulations of measurement data. This is the province of the section of Hydrology and Statistics so we returned to Mr. Ahuja. Our request was to see some actual data.

We were treated most kindly and he asked us to have tea at his home. Mr. Ahuja, who is from the Punjab, is rather light colored relative to many Indians. His wife, also from Punjab, is particularly fair with gray eyes. They lived in a flat furnished by the Indian government for civil servants. He explained that his job allowed a better house but the nicest ones are hard to get because of priority assignments. There is a long waiting list.

At the office Mr. Ahuja answered our request by assuring us that it could easily be done, but such details were not in his office. He pressed the small bell, just like the one on the desk of Dr. Rao, and tersely told a servant with a pink turban to take us to another office where we could be helped. The next office we saw was smaller but had a desk covered with green felt and a bell, yet quite barren of papers. This gentleman was kind and helpful but the records we wanted were elsewhere. In the succeeding office, quite like the others, we learned something. Mr. Krishnamurthy really was knowledgeable and helpful. He explained some details about how data were compiled and how analyses were made, but not in Delhi. The center of record keeping for the state of Bihar was in Patna, a city on the Ganges about 550 miles to the east. By this time we were not going to be further deterred.

The plane to Patna stopped briefly at Lucknow and Benares where the river is an important part of the religious life. We began to understand in a distant way how the rivers have become an integral part of everyday living. From the air most of the big tributaries of the Ganges are seen to be divided into several channels, and at low flow characterized by wide bars and islands of white sand. We were met at the airport by two personable engineers, Mr. Bannerjee and Mr. Shivalingam, both from the state office, known locally as the *Secretariat.*

We were taken to the government rest house, or hotel, called the *Circuit House*, a nice building with rooms opening on a portico. The bathroom has a cement floor, no toilet paper, soap, or tub, but a concrete shower that apparently discharges onto the backyard. The place was quite acceptable and, to our surprise, clean. Each bed was supplied with a mosquito netting that hung from the ceiling and covered the person once he is in bed. We were instructed not to get out of bed without carefully inspecting the floor, for cobras came in through the shower exit to sleep on the cool concrete floor.

With the mosquito bar over the bed I slept fairly well for it was warm but not hot. Because the toilet does not flush, one fills a bucket of water from the tap and dumps it into the toilet bowl. At Circuit House, the meals were good, defined as including a reasonable number of things one can eat.

At the Secretariat, we were introduced to the Chief Engineer, Mr. Mathrani, with whom the conversation quickly turned to the subject of data. Interestingly, no one there had been informed just what we were here to do, or even why we were here. Previous visitors had apparently been concerned only with a tour of the flood-control works. When we explained to Mathrani that we wanted to see data, he saw that we wanted to see the river and the data being collected, and he immediately shifted the contemplated schedule. We were to have left by train for a six-hour ride to Muzzafapur, but he arranged for us to fly there in half an hour the next day. Mathrani was fascinated to hear about our work in a small flume at Cal. Tech. and he did not know that channel bed configuration controlled the hydraulic roughness more than the particle size of the bed material. We had a long discussion about the effect of clear water released from a high dam and its tendency to erode the streambed downstream.

On the way to Muzzafapur we flew over the great Ganges and marveled at the vast expanse of water. Soon we landed at a strip of grass where there was no building, but we were met by the local engineer, Mr. Chatterjee, an eloquent and forceful man whom we liked very much. At Circuit House there, we were assigned a pair of rooms with no mosquito bar and a shower stall teeming with Anopheles mosquitos. There was no soap, no towel, and the usual type of toilet. That night we opened some of the bug bombs left over from army days and killed so many mosquitoes the floor became slippery.

The personal plane of the governor was sent to take us over the area to look at the rivers and the Kosi project. The pilot landed at the airport about six miles away, but that did not satisfy Chatterjee, who asked that the plane land on the cricket field behind Circuit

House. On the grass of the field the plane landed smoothly. It turned out to be a new Beechcraft Bonanza piloted by a debonair fellow who is the personal pilot of the governor, the chief minister of Bihar. We flew up the Bagmati, a large tributary of the Ganges, following it up to the foothills of the Himalayas. The hills are at least 3000 feet high, consisting of sediments dipping to the north and exposing large faces readily undercut by the river, thus supplying immense amounts of sediment to the flow. Some interior valleys in these hills were flat with sandy stream channels.

There was great change in stream pattern as the rivers left the hills and emerged on to the Gangetic plain. Closest to the hills, the channels were wide and braided but without definite width or high banks. Within a short distance they became relatively narrow and meandering, moderately incised. The whole flood plain of the Ganges seemed covered with fields and dotted with thatch-roofed villages. The arrangements of the fields showed clearly old meander scars that had been silted up full and level.

We flew NE toward the Kosi and followed the channels some distance to the place near the site of a barrage or dam. There a new small village had been constructed at a place called *Birpur* where a level piece of grass had been marked out for a landing strip. As we approached, I could see a swarm of people rushing from the village to the airstrip. Some two hundred people surged up to see the airplane and the visitors. We were led to a little house where tea was served and where we discussed the layout of the project being built. We had a chance to see the ongoing construction. The levee was being built with a cross-section like a pyramid with the top sliced off. The base measured about 80 feet, the top surface 15 feet above the base, and some 20 feet broad.

To outline the shape, two tall poles were erected, the tops of which marked the edges of the flat surface, and a string outlined the whole cross-section. This was a practical way of showing uneducated workers the dimensions of the thing they were building. The center of the structure consisted of a cutoff wall made of heavier, more clayey dirt. All was tamped by hand. This clay material was dug some distance away and carried to the site in an open cart to the place it was needed. The rest of the soil was dug with a shovel, put in a flat basket about three feet wide, carried on the head from one man to another until it was dumped in place.

Such a long levee passed several low spots, swampy areas called *chars*. To cross these, the levee was built with a gate of concrete or brick to permit water to flow through when the river was not in flood. Because there is no rock in the Gangetic plain, aggregate was made of brick broken up by a worker with a hammer.

We now were more than anxious to see some data, but Mr. Chatterjee ruefully explained that the records were kept at Sitmarhi near where the measurements were made. This was some hours away by jeep. The road was a one-lane track along the top of a levee that had rather steep slopes on both sides. There were hundreds of bullock, carts, bicycles, goats, and people who had to be dodged as the horn blew continually. At Sitmarhi our driver, Mr. Singh, was well known and everyone jumped at his directions, but the local engineer was out in the field and all the records were locked up. So that was the end of our quest.

The villages we went through consist of a series of buildings surrounding open courtyards, though many of these cannot be seen from the road but are obvious when seen from the air. The buildings are of sunbaked brick like adobe. Doorways generally are covered with a mat of woven strips from a palm-like plant. Transport is strictly by bullock cart or by carrying a load on the head. Even the smallest parcel is on the head and I never saw one fall off. In the villages there are many banyan and monkeypod trees, but they are not common on the flat plain that is divided into plots, each surrounded by a low embankment to pond irrigation water. Threshing is by a bullock walking around a central post to break up the grains. Winnowing of the chaff is accomplished by pouring the grain out of a basket in the wind. Clothes are washed in the many open waterholes or embanked ponds. These tend to be green with algae and generally look quite unappetizing.

As we passed one of these open ponds, one fellow was washing his water buffalo while other animals were submerged with nothing but a nose out of water. Many people were fishing using a little stick for a bobber. A woman was discreetly bathing, soaping her legs while sitting on a rock in the pond, completely draped in a sari, including her head.

All clothes seem to be originally white but presently are a dirty brown. They wash the clothes often but the water is nearly always muddy from buffalo wallowing in it. Slapping the clothes again and again on a rock will displace some of the silt but cannot get rid of all of it. The clothes are then laid out on the ground to dry.

When we returned to Muzzafapur, we made a courtesy call on the long-retired dean of the engineering group in that part of the country and a member of the prestigious Indian Civil Service. We told him of our quest for some original river measurements. He listened carefully and then explained.

"You don't seem to understand that there are no measurement data. Yes, they read the staff gage to see the height of the water, but then they pick up a buffalo chip, throw it in the water, count one to ten, guessing the distance. That is a measurement."

To return to Delhi from Patna, a driver was sent to Circuit House to take us to the plane. Shivalingam came to say goodbye. I took a picture of the engineer, the cook and the sweepers, who were delighted to be included. There was a long wait at the airport for the plane so there was a chance to watch other passengers as they checked in. There was an older man in a business suit and an American hat carrying a fancy umbrella. He was on his way to Katmandu with an entourage consisting of a fat wife, big as a horse, with slicked-back hair, lots of rouge, draped in a gold-encrusted wine-colored sari. Their daughter-in-law was even more gaudy in a lemon-yellow sari and gold slippers. She had strongly Mongolian features, a protruding chin, a flat nose beneath a receding forehead, spit curls over the temple, lots of surma (eye black) under almond eyes, slicked black hair set in combs and embroidery ties. There were a couple of insignificant grandmothers, a son in American clothes, three servants—a Nepalese major-domo in a poorly cut long coat, jodhpurs, and a round Nepalese black skull cap and two squatting women with tile (nose jewelry), one with a parrot in a cage. The baggage included iron boxes, baskets of stuff, peacock-tail fans, a wicker cote of live pigeons, pigeon feed in sacks, water bottles, food, trunks, guns, and ammunition. I watched the major domo peeling off bills to pay for 800 pounds of excess baggage as it was weighed, tabulated, and tagged.

Again, in Patna, I was asked to have a short private conversation with the governor of the state who kindly inquired whether we were satisfied with our visit. He had ordered his airplane to be put at our disposal, a favor gratefully received. Finally, he got to the main problem that bothered him.

"As you see," he said, "the river has many channels and the whole width, including all the channels, is great, so we are putting the levees nine miles apart. There are people living in that area and they might be flooded because they are between the levees. Do you think," he continued, "that my government has a responsibility to these people?"

"Once the levees are completed," I said, "then without a doubt they will be inundated in a flood. You certainly have a responsibility to them."

"Ah, you don't seem to understand. There are a million people in that group."

Returning to Delhi, we had a final session with Shri Konwar Sain, who made some pointed suggestions about some of the wording in our report. They were not matters of principle to us so we changed our statements without objection.

In saying goodbye, I had a last discussion with Mr. Ahuja about data. The values of streamflow they use and publish list the values to three significant figures past the decimal point. In teaching hydrology at the university I always explain that no parameters measured in hydrology have a precision of three significant figures. In asking Ahuja about this matter I said to him, "I saw several of the river staff gages pushed over by the flood waters, and therefore they cannot possibly provide a useful basis for discharge, and certainly not at the precision stated."

"Oh," he said, "I've just fixed that."

"Fixed it?" I said.

"Yes, yesterday I wrote a memorandum."

Problems of Development

Against this backdrop and some of the problems in rational water development, a few bald facts might be considered. Among the three countries of the world that have the largest discharges of organic water pollutants into their water bodies, the order, respectively, is China, United States, and India. In contrast to the engineering capabilities for control, the fact that India and the United States are together in pollutants speaks volumes about causes of degradation of a vital resource. Our country, by far the most capable and advanced in technology, not only fails miserably to husband its wealth, but promotes by actions at the highest level of government for the dissolution of our heritage. The administration of President Bush surpasses all others in the history of the world in actively promoting short-term profits gained from the dissipation of its wealth.

In brief, the adversaries in the field of water in America are the conflicting demands of large-scale agriculture and the needs of the urban population. But neither the president nor the Congress seems willing to judiciously balance these justifiable needs.

Looking at the whole world, there are more than one billion people lacking access to safe drinking water. Most discussions of water problems in the world take the form of elaboration of the kind of statistics quoted earlier. I chose to approach the problem in human terms, stressing the difficulties at all levels in the planning, the design, and the execution of projects for water utilization, even with the best of intentions of the leadership. These involve the physical character of the landscape, the effect of geography, of culture, of scientific knowledge, and, of course, personal preferences.

It is often the simple and inexpensive things given to poor communities abroad that both reach the people and are most appreciated. I met an educated Sikh in India who was born in a small village where everyone depends on the fruits of his own harvest. Thinking to help these people in their farming the US Bureau of Reclamation sent them a couple of tractors. My friend said, "Within a week of their arrival, the women in the village were wearing the piston rings on their ears."

On nearly all the continents of the planet, an appreciable portion of the country is desertic in character, arid or semi-arid. There is a general failure to realize that in such conditions, the profligate use of water to provide comfort rather than logic, greenery rather than stability, lawns and swimming pools rather than native plants, and the cultivation of crops that are in excess rather than scarce, is unrealistic. In America, the political influence of big corporations in opposition to the small farmer or worker continues to permit or even promote use of water inappropriate to long-term stability in the arid zone. The problems go far beyond the use of water; they involve immigration status and policy that ought to be influenced by climate and seasonal patterns of plant growth.

These types of conflicts exist at global as well as local levels. There is an imposition on the commons such as competition for extraction of water bodies that traverse two or more countries. Examples are many. The Salween and Mekong rise in the highlands of Tibet flow through China, Cambodia, Laos, and Vietnam. The Tigris and Euphrates rise in Turkey but are a principal source of water for Iraq. The Rio Grande begins in the United States, is gradually depleted and degraded in quality and quantity as it flows to the border and enters Mexico. Important tributaries of the Jordan rise in Lebanon, but downstream it is controlled by Israel yet is vital to the Palestinians.

I, among others, have forecast that the increasing pressure for control of limited supplies of water is leading inexorably to conflict among segments of a population, between states and between countries, and it will be the pretext if not the cause of war. This premonition has taken on increasing relevance in the last decade as weather phenomena have taken on a more menacing role. Even as far back as 1957 I gave a lecture at the Scripps Institution at La Jolla entitled "Water Problems in the Present Trend Toward Greater Aridity," and with the present global warming, problems are becoming visible nearly everywhere. It behooves societies throughout the world to take action to ameliorate the consequences clearly seen on the horizon.

While I had been busy writing a book in flood control, John Miller had done some innovative work on the nature of channels in high mountains. Because he had napped the geology of an important part of the Sangre de Cristo Range in New Mexico, he was well positioned to study the relation of bedrock type to the channels that develop there. He was also concerned with the nature of the dissolved solids emanating from these different geologic areas.

We had spent so much time observing the ephemeral channels of the semiarid areas, we felt it was time to study in quantitative detail the actual processes of erosion in that relatively barren region and measure the flows that seldom occur but still seem to accomplish much work. Our plan was to go to New Mexico at the start of the rainy season about July 15, camp out on a hill, and watch the whole horizon for thunderstorms that are so common in the region but seldom strike in the same spot. We would chase the storm wherever it was occurring.

The plan worked out very well and we were successful in getting to several good storms while the action was actually happening. John Miller and I spent weeks camping on the top of a low hill near Santa Fe, New Mexico, always looking for thunderstorm clouds anywhere within a twenty-mile radius. When we saw a cloud that promised rain we would drive madly toward it and succeeded in finding arroyos flowing enough times to compile a reasonable set of measurements. In such an event on arrival at the stream, we made measurements of the discharge, recording the width, depth, and velocity during the time the flow lasted, and took sediment samples at frequent intervals. Thus we knew from our samples that most of the sediment carried by these channels was sand with an additive of silt, and rocks in motion occurred in some but not all of the channels. It was known also that during such a flow event, the channel bed was eroded to some depth but tended to redeposit at the end of the flow. Such known scour had never been measured in an organized way.

Many interesting things were observed, including successive wave fronts coming down the channel at definite successive intervals of time. We measured the area covered by storms, and the relation of tributary contribution to the channel flow. During the nonstorm periods we studied the channel network and compared it with that known from other climates and topography. Our published paper still stands as one of the very few that includes on-the-spot measurements of flow parameters in desert streams.

That summer was a satisfying start and provided impetus for a more extensive investigation of erosion processes in the same kind of channels. Interest in the latter was enhanced by the knowledge that these ephemeral channels do not exhibit the pool and riffle sequence so prominent in gravel-bed streams elsewhere.

The ubiquitous occurrence of bars of gravel in rivers has been known for a very long time but seldom studied in detail. The point bar on the inside of a meander curve is a prominent phenomenon but the surprising fact is that bars occur in straight reaches of channel, alternating from one side of the channel to the other side.

I had shown that their spacing was consistently equal to about 5 to 7 times the width of the river. The same was discovered independently by the talented icthyologist and ornithologist Tom Stuart, a scientist at Pitlochry, Scotland. He showed me a Scottish river that was being straightened by a dragline. He had approached the machine operator and asked him merely to put a bucket of gravel opposite each stake that Tom had put along the bank, 7 widths apart. Within a year those piles of gravel had been smoothed out by the river flow and now were stable gravel bars.

To start an investigation of how a bar is maintained, we placed painted rocks on bars and observed that at moderate flows, the rocks moved downstream but appeared to be replaced by other rocks so the shape and size of the bar did not appear to change.

Recently McBain and Trush of Arcata, California, have shown that for stream health, it is essential that a flood of 5- to 15-year frequency move all the rocks in the bars and reset or reform their distribution, and that remolding of the bed resembles the one that was washed away. But there must be gravel contributed by the tributaries to provide the replacement material.

Because gravel bars were so ubiquitous in rivers we had studied, it was striking that these were missing in the semi-arid valleys in the southwestern states that had dry, sandy ephemeral channels, ones having flowing water only a few times a year and each occurrence lasting only a few hours. This seemed common in the semiarid portion of the country and needed explanation.

We had gained experience in the study of these ephemeral flows. Thus we knew from our samples that most of the sediment carried by these channels was sand with an additive of silt, and rocks in motion occurred in some but not all of the channels. It was known also that during such a flow event, the channel bed was eroded to some depth but tended to redeposit at the end of the flow to the same level it was before the storm.

Such known scour had never been measured in an organized way. Further, the actual origin of the sand that moved had never been studied in detail. All one knew was that the hill slopes eroded gradually, hills and some gullies were cut by rain and runoff, but how much, where, and with what result was unstudied.

Aimed at construction of a sediment budget showing source and distribution, we planned a variety of measurements, using techniques we invented and thus were new and untested. They included the following, which will just be mentioned and then described.

1. Measurement of depth of scour using scour chains
2. Survey of the surface as a cross-section across the channel and up on the side slopes
3. Nail and washer plots to measure sheet erosion
4. Headcut form, height, and rate of retreat
5. Resurvey of deposits of sediment in a small reservoir
6. Painted rocks to record when they moved, when and where they were deposited
7. Recording the alluvial sequence of valley sedimentation and erosion by mapping of terraces and their stratigraphy

Plans Put on Hold

This was planned as a long-term project so the planning and preliminaries were extensive and methodical. The work began in 1957 and lasted about 10 years. But just as we were getting under way, both of us were shaken by big changes in our lives. I had been asked by the Director of the Geological Survey to assume the job of Assistant Chief Hydraulic Engineer. To be my partner, he chose a groundwater hydrologist from Idaho, Raymond L. Nace, as another Assistant Chief. My field was to be in control of the budget, Ray was to be in charge of operations. These were big assignments in an operation with more than 300 offices around the world and more than 3000 people.

At nearly the same time, John Miller had just as large a shock when he was offered the professorship of geomorphology at Harvard. We both vowed that despite the heavy workloads, our research project would continue.

We chose a watershed for study located a few miles out of Santa Fe as well as work in a nearby basin. Arroyo de las Frijoles is an ephemeral basin, our study area being 7 miles long and one mile wide. It eventually drains into the Santa Fe River. The first task was to make a traverse by transit and tape, establishing turning points marked by 3-foot steel pins driven into the ground. These were then surveyed by level to establish elevations for each. We then began a map by plane table and alidade recording the nature of the channel as we went. There were a few cobble-size rocks scattered at random and partly buried in the sand. They did not constitute a bar.

One of the advantages of mapping is the fact that the process makes one look in detail and it soon became apparent that there was a collection of pebbles, 1 to 2 inches in size, on the surface and on alternate sides of the channel, usually spaced 4 to 6 channel widths apart. The pebbles of these bars are strewn as a veneer, never more than a few grain diameters deep. So these were the substitutes for gravel bars that occur elsewhere.

We then established a series of cross-sections of the channel for resurvey, monumented at each end by a long steel post extending above the surface about 6 inches. The elevation of each was determined by a level survey.

To measure the depth of scour of the channel during a storm flow we developed a technique that differed from those used by other investigators and was far more satisfactory. Each observation point consisted of a chain about 4 feet long having links with openings of $1/2$ and $3/4$ inch. At the end of the chain a cobble was wired on to anchor it. A vertical hole was dug in the channel about 5 feet deep, a chain with a cobble was lowered into the hole, held vertical while the hole was backfilled. The extra length lay on the surface of the bed. The location of the chain in the channel was carefully measured. When a flow occurred, the chain would be forced to lie flat on the top of the unscoured surface, and, as usual, the sand redeposited to the same surface elevation that existed before the flow. After the flow the location was found again by survey.

Each cross-section had 8 chains installed and chain sections were distributed along the whole length of channel. There were 90 chains installed. We found that the depth of scour increased with the peak discharge. For modest flows the depth was 0.2 or 0.3 feet, but in large flows, 0.8 to 1 foot.

In digging the holes for installation of the chains, we practically never encountered a cobble-size rock. Yet, as mentioned, such rocks were seen randomly scattered along the channel and exposed at the surface. Considering the measured scour, one would expect these rocks to be lying at the depth of scour, but rocks are found at the surface or slightly protruding. The explanation apparently is that they are pushed to the surface by the dispersed grains when all grains in the scour depth are in motion. Brig. Bagnold had shown that stress by dispersed grains exerts greatest pressure on large particles and thus they are pushed to the zone of zero stress, which is the surface. These are useful and interesting parts of the puzzle.

Nails in the Ground and Tragedy

The measurement of erosion and deposition in an arid climate stream are more complete in this investigation than any other in the published literature. Particularly important was our introduction of the nail-and-washer method of measuring sheet erosion. A metal nail, 10 inches long with a washer, was driven into the ground until the washer and head were at the ground surface. As erosion occurred the nail gradually protruded above the surface and the washer sank down to the new surface. Measurement of the gap between the head and washer was the erosion during the period between measurements. A total of 183 nails with washers were installed and measured nearly annually. To measure erosion at a nail we knelt near it and with a millimeter scale read the distance from head to washer. These measurements had been made in several years and the net result of these unique measurements was very satisfying, but it came at a great cost.

In the summer of 1961 on a particular Friday several years after the project began, John and I had spent a morning on data collection and ate a modest lunch in the shade of a piñon tree. We wrote out an outline of the book on geomorphology we were about to begin writing. The chapters planned were divided so we knew which parts each was to write. We went back to work measuring the nail-washer pins in Coyote C Arroyo, so named because of the many signs of coyotes there. At the end of the day I drove him to the airport and he flew to Boston.

Unbeknownst to us at the time, he had been bitten by a flea as we worked kneeling on the ground. He came down with a serious illness that was new to the physicians in a small hospital near Boston so it was not initially diagnosed. John died of bubonic plague within three days. I learned later that plague was endemic in a few places in New Mexico and this turned out to be among them.

Pursuit of Our Aims

The loss of John made me even more determined to carry on this study and to write the book we had planned. For the next couple of years many evenings were spent compiling, reading, organizing, and writing but the result was successful and the book, *Fluvial Processes in Geomorphology*, was published in 1964 and was well accepted in the scientific community.

The reasoning John and I used to design our project came from various sources. It had been observed by river engineers that rocks tend to hide behind one another and a partly hidden rock does not get the same downstream directed force as would a rock of similar size unprotected by its neighbors. This indication of interaction among rocks led us to search the literature for analyses and they were not hard to find. Highway engineers had studied the movement of traffic and described in mathematical terms what every driver knows from experience. The closer your car is to the one in front of you, the slower you must go. Rules of the road have been developed stating how many car lengths you must leave between you and the next car as a function of your speed. This literature showed what we recognized from driving experience: that cars proceed in platoons. They are not randomly distributed along a highway. Between groups of cars or platoons there would be an open gap, relatively free of cars. These distribution patterns are due to the natural way a driver uses his brake. He slows down when he gets closer to the car in front.

All these observations led us to the conclusion that rocks on a streambed interact with each other so that the individuals in a group, being harder to move than if alone, tend to bunch together so a gravel bar is related to a platoon of cars. Here was a concept and we looked for a way to prove it.

John and I both felt that just stating this to be a fact or to have merely a laboratory study to prove it would not convince the field geologist so we must devise a way to investigate the idea in a real stream in the field. We decided to place rocks of known size on a streambed at different distances from one another to see if those in close proximity to its neighbors would be harder to move than if they were spaced far apart. To start the experiment we walked along the Arroyo Frijoles collecting cobble-size rocks that were sparsely scattered. They were taken to a central place, individually painted, weighed on a scale and had the weight in grams painted on each. They were divided into six classes of size, the average weight ranged from the smallest, 400 grams 65mm diameter, to the largest, 9000 grams 230 mm diameter. These painted rocks were placed on the streambed in groups of 24 rocks, each group including four rocks of each of the six weights. The rocks in a group were spaced at certain distances one from another, the closest spacing was 1 diameter (the adjacent rocks touching each other), to 8 diameters.

Flows in such an arroyo occur on the average of 3 times each summer. After each flow, our colleagues in the Santa Fe office of the Geological Survey went to the field and computed from survey of the water-deposited debris, the peak discharge of the event. After each major event, John, Bill Emmett, and I flew to New Mexico to walk the full length of the project, picking up each painted rock and carrying it back to its original place on the bed, recording the distance traveled and other features. Over the course of this project we monitored 14,000 painted rocks, including those that did not move as well as those that moved.

At the end of the years of study our report confirmed the hypothesis that rocks close together are more difficult to move than when widely separated. Laboratory experiments and these field studies indicated that rocks in a stream travel in platoons like automobiles. Gravel bars as a whole are relatively stable and the location moves but little though the individual rocks composing the bar, arrive, stay a period, then are moved on. Thus a gravel bar is a kinetic wave made up of individual items that discontinuously move. No other theory of gravel bars has been published and this theory has yet to be discredited.

The study of erosion rates showed that by far the greater part of the sediment derived from the watershed came from the surface erosion and not from the spectacular gullies that look so erodible.

The nail-and-washer data proved to be invaluable for the sediment budget. The study produced a large number of observations of the distances that individual rocks moved during a runoff event. Intuitively one might suppose that small rocks move much farther than large ones but this was not the case in this study for the distances moved were random and not related to size. The farthest that an individual rock moved in a single excursion was about two miles.

This study of erosion and sedimentation showed that many of the most elusive problems in physical geography do not lend themselves to models prepared on computers because field measurements of the requisite parameters are not available. Field measurement follows from detailed observation and a subsequent plan for organizing all aspects of the investigation.

5
Science as a Craft

At this point in my career when I had a large responsibility in administration and was starting a research program in an organization that never had had one, it was time to review where I was and what my relation to science itself was. My closest friend and colleague was gone and I had to decide where I was and where I should go next.

The first field season after university training I spent with John Miller in eastern Wyoming, a time for applying the field knowledge acquired under classroom conditions. This season and several years following were spent in traveling and recording observations, periods of time some onlooker might suppose were just having a good time, camping, fishing, climbing, and hiking. Perhaps that is the reason young geologists enjoy themselves, and indeed, some critics apparently thought one should not enjoy work in the field. A beginner is learning what to look at, how to record it, and gradually how to interpret it. These certainly were the steps I and many of my colleagues were taking. In my case, the principal result was the formulation of questions, for answers may follow from posing problems to solve. I learned a lot from John Miller, especially glacial geomorphology and soil chemistry. No other person provided me the opportunity to discuss in detail the whole gamut of geomorphic problems and field conditions. Together we became good surveyors and increasingly experienced in organizing our material for scientific reports.

The channels we saw and measured in the Powder River basin were wide and shallow. The banks were mostly silt and the streambeds had but little gravel. My earlier experience was with trout streams that were generally bedded with gravel. In contrast, the arroyos I had studied were like the Powder River for reasons I did not understand, so after the first seasons I chose the problem of channel width. To approach this I needed to study a variety of rivers, more than I could visit in the field, so I looked for any compilations that might have been made.

Framing the Questions

The hydraulic geometry showed me an important lesson, that the framing of a question is vital to progress in research. This, then, became the beginning of extensive investigations of the shapes of river channels in different environments, a subject that has engaged many workers for a long time. Throughout these first years I was beginning to formulate a pattern or a sequence of moves in research. It is necessary to state a problem and a question that it presents. Such a pattern leads to a series of observations of those parameters that seem involved in the measurement data. The statement of question, if framed correctly, is a hypothesis. In this case, the statement was as follows: Channel width is related to some combination of other hydraulic parameters such as discharge, velocity, depth, and slope. Because large rivers are wider than small ones, there must be a relation to river size and that must be a function of drainage area. It also follows that pattern, shape, flow regimen, overflow events, and relation to the riparian vegetation must all be interacting and interdependent.

In this I had stated a hypothesis and had to discern a procedure to examine the problem in a formal way. All I know is that the choice of a problem was a useful first step.

Had I appreciated better the extensive literature on research philosophy, perhaps I could have seen a logical rather than merely an intuitive approach. But I did know that inductive reasoning came from careful observation and thus measurement. Yet logical instinct led me to think that investigation begins with careful collection of observations and facts. In this I became increasingly conscious of the wisdom of Dr. Eugene Auchter, Director of the Pineapple Research Institute. His instruction to me as I entered his employment was definite; do not do anything, observe, record, travel, and learn from others. This was the fulfillment of what I had learned from experience under various supervisors. The ideal attitude toward an employee is to provide freedom, support, and encouragement. In this review I studied in depth the instruction of President Jefferson to Meriwether Lewis that guided the famous expedition:

> you will take careful observations . . . at all remarkable points on the river, at rapids, at islands, and other places and objects distinguished by such natural marks and characters of a durable kind, as that they may with certainty be recognized hereafter.
>
> Your observations are to be taken with great pains and accuracy . . . to be entered distinctly and intelligently for others as well as for yourself.

Jefferson continued:

Other objects worthy of notice will be
the soil and the country
the animals of the country generally...
the mineral production of every kind...
volcanic eruptions;
 climate, as characterized by the thermometer...
the dates at which particular plants put forth or lose their flowers
 ... times of appearance of particular birds, reptiles, or insects.

The whole letter of instruction reminds me that Dr. Auchter's instructions to me were of the same tone: observe and record. These thoughts came to my mind often and in various circumstances when the purpose of my travels sometimes became dim. Most important, the review heartened me because for years I had adopted some of the particular practices that Auchter and Jefferson had in mind. I had been keeping a journal comparable to that kept by my father, recording in some detail the happenings and observations on each trip both short and extended.

This type of journal is not a diary because it is not kept daily but on trips it contains far more detail that then would be usual in a diary. The record began in earnest in 1932 and continues to the date of the present writings.

The details in Thomas Jefferson's instructions are particularly interesting because without such prompting I had been recording many of the details he mentioned. Of particular interest was the recommendation of phenological records, for just such a detailed record was kept by Aldo Leopold at our small farm in Wisconsin and continued after his death by his daughter Nina, her husband, and his son Carl. It now spans 60 plus years and Nina observed 180 items in her list of observed phenomena.

In looking at a scientific problem, clearly the approach I have taken is the inductive method using a collection of facts observed in nature to propose a hypothesis or tentative conclusion. It seems to be the logical extension of the procedure following the instructions of Auchter and Jefferson. Once a question is posed or a hypothesis is stated, then the approach to a solution or an advance toward a solution can be assisted by viewing the problems in terms of a craft. In any craft there are objectives, raw materials, tools, techniques, and products. There are several aspects of any craft that are important to mention. The techniques can be taught but using them requires care, patience, persistence, and practice. Some persons become more skillful than others.

Very important is the aesthetic pleasure that a practitioner of a craft may take in a final product of which he can take some pride. Striving for such satisfaction can be a source of continual stimulation. It seems to me unfortunate that in teaching people who aspire to a career in science there is usually little if any discussion of techniques. Perhaps it looks demeaning to talk about how to do science and the techniques useful in approaching a problem. Yet there are well-known scientists who have done just that in one way or another. Bacon introduced the idea of multiple hypotheses. Charles Darwin in his autobiography gave some details on how he operated in his massive output of scientific material. In doing so he was passing on to us valuable lessons in how science was approached by one of the best. He wrote:

> I think that I have become a little more skilful in guessing right explanations and in devising experimental tests; but this may probably be the result of mere practice, and of a larger store of knowledge. I have as much difficulty as ever in expressing myself clearly and concisely; and this difficulty has caused me a very great loss of time; but it has had the compensating advantage of forcing me to think long and intently about every sentence, and thus I have been often led to see errors in reasoning and in my own observations or those of others.
>
> There seems to be a sort of fatality in my mind leading me to put at first my statement and proposition in a wrong or awkward form. Formerly I used to think about my sentences before writing them down; but for several years I have found that it saves time to scribble in a vile hand whole pages as quickly as I possibly can, contracting half the words; and then correct deliberately. Sentences thus scribbled down are often better ones.
>
> Having said this much about my manner of writing, I will add that with my larger books I spend a good deal of time over the general arrangement of the matter. I first make the rudest outline in two or three pages, and then a larger one in several pages, a few words or one word standing for a whole discussion or series of facts. I find I never make a detailed outline and then expand it. Often a whole paper begins with a few sentences written down. During the actual writing I find I am constantly changing the outline. As in several of my books, facts observed by others have been very extensively used. I have always had several quite distinct subjects under study at the same

time. I may mention that I keep many large boxes placed in cabinets with labeled shelves, into which I can at once put things I have cut out from reading that I think I might use in the future.

We can look back at the great Darwin and take some solace in the fact that even he sometimes had a hard time getting thoughts down on paper in a way that satisfied him. I have had many students who claimed that writing was hard work and took a great deal of effort. Kirk Bryan of Harvard, a very successful teacher, said to graduate students who were having a difficult time writing, "The way to write a scientific paper is to write down one word after another. He said this with a wide smile but there is a lot of good sense in his words.

Though I did not know what Darwin did, I never wrote a paper by expanding an outline. I wrote books from an outline of just one page. I have found it helpful to advise the following as a technique to help writing: Arrange the figures, illustrations, or maps you have in mind in the order in which they will be introduced. Then write a description and discussion of each. In doing so the elements of a paper will have been written.

The well-known geologist of the University of California, Andrew C. Lawson, wrote extensively on the spirit and method of science, pointing out that, before the advent of modern science, thinking men used what is known as the deductive method, that is arguing from known principles to deduce process or cause of individual cases. Modern procedure in science generally uses the inductive method, drawing on a large body of observed fact to draw a hypothesis.

Interestingly, Lawson did not include any detail of his own procedure of keeping notes, recording observations, or compiling descriptions. I have found it useful or even necessary to discuss these details as part of instructing students. Some students and young colleagues have said that these seemingly mundane procedures regarding note keeping and writing have been the most important results of my teaching.

A useful summary of how to approach problems in science was discussed by Karl Popper, who proposed that a hypothesis could be proven false by an experiment. The same idea was made more understandable in the paper by John R. Platt, "Strong Influences," published in *Science*, 1964. After discussing the advantage of stating multiple hypotheses, his advice is to ask the question, "What experiment could disprove your hypothesis?" The method of strong inference, he says, suggests we, "write down our alternatives and crucial experiment, focusing on the exclusion of a hypothesis." How can the hypothesis be disproven?

My own experience dealt to a great extent with geomorphic processes and the effect of rainfall on the landscape, the collection of their widely dispersed units of runoff, surface runoff, collection in the rills. Rills expand into larger units that take on the forms and characteristics of channels in the movement of water and eroded products, that is sediment derived from the gradual decomposition of earth materials.

A natural consequence of such activity is to notice processes, actions, or results that are not immediately explained and thus there develops in the mind a question inviting answer. Usually a tentative hypothesis or more than one is formulated and this then becomes a subject to be explored by reading, cataloging, and possibly experimentation.

There are many examples in the geological and geophysical science in which initial concepts, though powerful and useful, have been modified or amended as more data and analysis developed. Thus hypotheses have been elaborated and alternative views resulted, in which the most valuable and general was the most simple.

Thinking of examples in the field of geomorphology to which alternative hypotheses might be enunciated, I find myself hard pressed to find more than a very few that can be described as having alternative hypotheses. Perhaps this is the result of the relative youth of this segment of scientific inquiry or for the limited number of specialists who have contributed to it.

At this time when I was entering that phase of life that might be called mid-career, I was able to articulate certain precepts that guided my sense of direction in science. Beyond the usual time and attention to the job of earning a living that always involves responsibility for management and administration, there is in everyone's life the choice of where to spend his time and effort. At least part of that choice facing one in investigation or research concerns how to choose a problem on which to work.

A Plan for the Work

The first precept guiding me is that an investigator should not confine himself to one problem. I tried to be engaged in three problems at a time that can be defined as:

a) Problems of fundamental nature that may be difficult or even insoluble but might, if successful, illuminate some important relationship or principle.
b) Problems that involved field observation that I defined as subjects that made you look.
c) Scientific problems that required time and effort that was plain enjoyable, camping, boating, hiking or sightseeing.

With regard to the first of these, it is recognized that such a project may take years of work, that is long term, not susceptible to immediate results.

The second category involves observation of field conditions, especially the group of which require mapping, by instrument or by field sketch maps. Mapping or just sketching is one of the most important skills in physical geography. Mapping forces one to look.

The third category is useful because it avoids pressure for results, but such activities have led to new places and to new ideas, and these in turn offer new perspectives.

At this point I want to say something about long-term projects. The projects mentioned as those of a fundamental nature can further be subdivided as follows:

Those the purpose of which is to keep observations, continuing through time to see how processes operate, and

Those for which a hypothesis has already been developed and the project is to measure parameters that may help support the hypothesis or find it lacking.

One might ask from where are these possible problems originating? My answer is from one's own experience. Most of my long-term projects came about from a series of events many years earlier. In the first decade of the century the great gullies that were cutting through the alluvial alleys of the southwest were still active, and there was much speculation about when and if the process would slow or cease. In 1927 Kirk Bryan had measured a cross-section of the Rio Puerco in New Mexico for the purpose of observing change over time, but in 1938 the monuments could not be found and the data were lost. To start again I ordered a survey team to measure a new cross-section in the vicinity of where Bryan had worked.

After the war I wished to examine that survey and see whatever change had occurred but was disappointed to find that the original notes were lost and the monuments could not be located. Also, I went to the see what was left of the 100-meter quadrat I had mapped in 1933. It was destroyed by grazing, the plants had been entirely eliminated by animals, and the maps I had made had disappeared.

With this history and because my network of surveys on the Watts Branch were to be used over a long period of time, I determined to establish a system for preservation of such records. At a meeting in Italy I announced the establishment of an international data repository called the *Vigil Network* for such data that should be preserved, two copies to be kept separately, one in Uppsala, Sweden, and the other in Washington, D.C. Scientists were invited to place copies of their data in the file for safe-keeping. Recently the file has been moved to the Hebrew University in Jerusalem, and Denver, Colorado, respectively. The file grew slowly but progressively. Osterkamp, Emmett, and I published a list of all the files on record and urged participation.

There is another source of ideas that is usually not recognized. It is pure inspiration. By that I mean the occurrence of an idea, a thought, a relationship, or a vision that had not been previously in mind in the recent past. Such an idea seems to occur without relation to previous considerations. That such a phenomenon is even mentioned may seem to some as unscientific and partly obscurism, yet it is a fact that ideas do surface in the mind in unforeseen ways. Such ideas should be immediately written down lest they be forgotten. In my lifetime I have experienced on occasion some new ideas that seemed to spring from nowhere but turned out to be worth exploring. Some other scientists say they also have had such inspirations.

I urged every one of my students to choose the name he or she wishes to appear as author of a publication. Preferably it should be the first name, middle initial if there is one, and the surname. Use this on all documents and do not vary from it. Two letters and a surname is an unfortunate choice of name on a publication.

Keep a file of copies or reprints of everything you have published and all important manuscripts, even if not published. When enough have been saved, construct an index. Arrange in chronological order, and have a volume bound in a strong buckram. Over time this will be the gross output of your professional life.

The keeping of notes in the field, office, or laboratory, is an important subject. At the very beginning of my work in geology, I saw the advantage of the custom in such fieldwork to carry the field notebook in a leather case on one's belt. I adopted this practice but went further. The field notebook must be sewn like a book, bound in stiff covers, and should always be the same size. Notes in loose leaf or notes bound in spiral binding should be avoided. All my students and colleagues have adopted this procedure and are glad they have. For years we bought what is called a *miner's fieldbook*, $4\,1/2 \times 7\,1/2$ inches in size, characterized by tabular columns on even-numbered pages, and cross-sectional paper on odd-numbered pages. This permits the right-hand pages to be useful for maps or for graphs as well as for tabulations. In my field books I always pasted in the end of the book graphs or diagrams that provided numerical values useful and summarized data. In recent years we prepared and had published a field book that included a wide variety of graphs, diagrams, and tables. This is called *The River Field Book*.

In addition to the field notes my procedure is to have a separate journal in which I record in the form of text, a history or account of each principal trip or outing. It is not a diary for it does not record day-by-day events but does cover field experiences in a narrative form. The journal is usually illustrated with sketches, maps, and photographs. These journals, kept during years of experience, in my case constitute 11 bound volumes, and the field notebooks now include 70 books. This is not a prescription for others, but simply a statement of the way my personal records have been kept.

Finally, I urge students to buy reprints of their publications at the time the page proof appears. Spend the time to develop a list of colleagues to whom he will mail reprints. A person who receives a reprint from the author is far more likely to read the paper than if it appeared in a journal that the recipient regularly scans. A list of 50 or more persons to whom a reprint should be sent is recommended.

Another precept concerns dealing with other persons in science. If I am asked to review a manuscript, I want my name to be included so that my comments are not anonymous. Further, I will not criticize another investigator in print. If I have a disagreement with him I will either spell it out in a letter to him or comment on the principles or methods without personal reference to an individual.

I repeat: Have several studies going on any one time.

These simple precepts have been useful guidelines for my conduct in affairs of science.

The Gates of Lodore

In Brown's Park on the banks of the Green River in 1962, I sat looking at the silty water of this wide, not very deep river, much larger than the creeks and streams on which I had been working. The small channels I had chosen for study could generally be waded with care even at high flow.

The main characteristics of rivers of all sizes could be seen in these natural but small-scale examples, but in the back of one's mind is always a question, do the great rivers of the world really act like the small ones? I wanted to get some feel for this question and I was now was engaged to do so. Having pored over the writings of those who had experienced the big channels, especially John Wesley Powell, G. H. Matthews and C. C. Inglis, I needed exposure to a new level of stream magnitude.

Powell started his expedition of 1869 at Green River Station when the railroad reached the river. He explained that the sixty-two miles before reaching the first big canyon presented merely some fast riffles but he wrote a "placid stream past the carved cliffs of the *mauraises terres*, now and then obtaining glimpses of distant mountains." But now they were faced with the narrow canyon bordered by cliffs a "five hundred—a thousand—fifteen hundred feet high." These cliffs of purple quartzite in the evening give a glow of color reflected off the water that can hardly be imagined. He named it *Flaming Gorge*. I now looked at the same imposing entrance and tried to imagine the feeling of those brave men who looked at it for the first time. A few years later a huge dam was built at this same place, flooding the hills and valleys for miles upstream.

But at this time, I estimated the river to be about 200-feet wide, probably 5-feet deep. It had many silt or fine sand bars, more lateral than central to the channels. At the stage I observed it, it was 4 to 5 feet below bankfull, judging by the many remnants or segments of a flood plain. In addition, there was a higher level terrace averaging 12 feet above present water level, or 7 feet about bankfull.

I had signed on with Buzz Hatch of Vernal, Utah, the most experienced riverman in the area, and at that time one of the few offering commercial rafting trips. Before dinner Bill Emmett, my colleague, and I climbed up about 700 feet on the cliff and sat on a quartzite promontory above camp when we could look both upstream and down. Powell and his men had done something similar 100 years ago, "a climb of more than a thousand feet to a point where we can see the stream sweeping in long beautiful curves through the gorge below."

It was my intention to measure, if possible, the successive depths of the smooth pools and the rapids reaches. To do so I intended to use the depth sounder that had recently been used by my associate Bob Meade, whom I had sent to Brazil to make the first-time measurement of the great Amazon, which he did successfully with the help of my friend Hilgard O'Reilly Sternberg, who at that time lived in Brazil. But I was not sanguine about whether the instrument would work under present river conditions, and I was not impressed with the condition of this boat to accommodate it.

The boat was finally launched about 2 p.m. It was a rubber pontoon 25 feet long. The equipment appeared to be in poor repair, as much due to lack of care as other causes. The air pump to fill it was with a pulley too large for the shaft and was shimmed with a tin can. We finally took it to the Ranger Station and drilled through both the pulley and shaft to keep it on with a bolt through the hole, but by this time the men had pumped up the boat with a hand pump. The pontoon itself was in rather poor shape. As an example, the rings for attaching ropes were torn off.

Immediately on entering the box canyon, there is a series of rapids that Powell felt required the boats to be lined, that is let down at the end of ropes, so that when a boat entered the usual eddy at the foot of the rapid it can again be boarded. Our boatman decided to run this reach, and though it was rough and we caught many waves, it was passed without incident. But presently we approached the place Powell called *Disaster Falls* because it had caused the wreck of one of his larger boats. It was usual in his case as in ours to land in the relatively smooth water above a major fall to scout the rapid and see the locations of the major rocks and the holes behind them, but in his case, one of his boats did not make the landing upstream and was carried into the V of fast water falling into the hole below. After the first fall of about ten feet the water tumbles down forty or fifty feet past big rocks and against one of this his boat, *The No Name*, crashed, throwing the two oarsmen into the maelstrom, from which they managed to land on an island at midstream. The boat was irretrievably lost but they did manage to recover some barometers that were in a compartment near the bow of the wreck.

Disaster Falls is the place the Ashley Party also was wrecked even before Powell arrived there. Powell chose the name *The Gate of Lodore* after a location in Cumberland, in England, which displays "waterfalls . . . between steep wooded hills."

Our boatman surveyed the fall and decided to run it. The boat swung out into the current and the boatman rowed it out toward the center down which we passed into the V of smooth water that nearly always occurs at the lip of the fall. Then one looks down into the dashing waves in, over, and beside the big rocks that usually are present. Below each large boulder is a hole that may be as much as 5 or more feet deep followed immediately by a tremendous wave, with big waves in sequence below. The strategy, of course is to miss the deep vortex behind the largest rocks and then to move laterally to avoid the big waves engendered by the main fall.

As we went over the lip of Disaster Fall, I could not in my limited experience judge how well this boat negotiated the rapid, but, for sure, the bow of the boat bent up about 40 degrees and hit the big wave that flooded the front of the craft but did not capsize it. This was the beginning of my experiences with the major rapids that I encountered.

After looking at the beginning of Hell's Half Mile, our boatman chose to run part of it with an oarsman from another party. He told the paying passengers to walk around the series of rapids. The other boat party lined their craft down the half mile of rapids. Powell and his party chose to portage and line boats down the rapids of Hell's Half Mile, a process that took him a couple of days.

We saw four groups of bighorn sheep in the canyon. One big ram with horns that curled into a full circle was escorting about four ewes. All three were grazing on the narrow sandy terrace remnant that stood some 10 feet above the river. I could not determine what they were eating. A track of a bighorn is longer than that of a deer, and parallel rather than pointed. The feces are slightly larger and ellipsoidal rather than round as in a deer.

On the second afternoon we reached a delightful tributary valley called *Jones Hole* coming in on the right bank. It was so pleasant we decided to spend a whole day there. It provided a chance to hike up the valley to see how the stream traverses the bedrock sills. In some places there is an impressive waterfall, while on others merely a very steep reach.

The day at Jones Hole also allowed time to examine some of the processes. In its lower end the creek flows in a narrow box canyon between limestone cliffs some 1500 feet high. The stream is some 20 feet wide, clear, with some beautiful pools. There is a gaging station about a mile above the mouth and there we made our standard measurement of bankfull stage, pebble count, lithology count, slope. I estimated discharge, for we did not have our regular instruments.

I rigged some flies on a willow stick and fished for a couple of hours without a strike. Most of the time was spent on the forms of slope seen on the canyon walls. The floodplain here is about a foot above medium stage of the creek, or 2 feet above the streambed.

Above that is a terrace at 8 feet, sandy in the upper 4 feet and gravelly below.

This small tributary had cut a deep and narrow canyon as it flowed to the big river. The upper half of the valley walls were vertical, about a thousand feet, cut in massive limestone above which was a thick sequence of buff-colored sandstone streaked with black manganese. The lower part was characterized by different forms depending on whether it was a prominent nose, or a re-entrant, that is an incipient valley. The base of each nose was covered with talus, convex in plain view, consisting of rocks of various sizes from angular blocks the size of a house to cobbles. The profile of the talus was straight and at an angle of repose of the rubble. The re-entrants, or coves, were in the shape of a fan consisting of water deposited silt and sand with irregular inclusions of boulders and blocks of rock.

The alluvial fill was trenched by the present stream that now flows within the walls of the cut it carved. The gullies on the fans have not formed a deposit on the alluvial surface of the terrace, and that indicates the gullies are older than the alluvial fill of the valley.

It appears then, that at some time in the past, probably in glacial times, the climate was periglacial, that is, cold, with freeze and thaw that caused rock blocks to be riven off the cliff forming talus slopes, but in the coves enough flow was concentrated that the riven rocks were carried away or deposited on a growing fan. Later, in a somewhat different climate, the fan was no longer building and was rilled by runoff. The fan consisted of small debris flows, leaving many local crescent-shaped lines of rock that marked the end of the debris flows. Later the climate changed again and the valley accumulates an alluvial fill.

During the end of the time when the valley was aggrading and sediment deposited on the valley floor, people were living in this little tributary. The buried hearths we found were at the edge of the valley under an overhanging cliff that provided some shelter to these inhabitants. Subsequently a change in climate caused a gully to form in the alluvial fill that grew in size and depth until the little stream was 8 feet lower than its previous position on top of the alluvium, at the level of the present stream. The level of its higher position is now marked by a terrace along the length of the valley.

From these relationships, it appears that the frost riving occurred during glacial time, followed by a period of valley aggradation seen in most valleys of the semi-arid west. The trenching of this deposit probably took place in the 19th century, and the Paleoindian occupation in the time of the Basketmakers, circa 800 AD. This is the type of reasoning that geomorphological observation permits one to make, conjectural but logical.

The party continued downstream to the mouth of the Yampa River that enters the Green in Echo Park, a landmark location in environmental activism. The Bureau of Reclamation had proposed a dam in Echo Park that was heavily attacked by many people led by the Sierra Club. The result was abandonment of the Echo Park site and a concentration on a downstream location, Glen Canyon, where the dam was finally built to the dismay of the many citizens who valued the natural river. Near the end of the canyon of Lodore the river goes through a final notch at Split Mountain, where there is one of the most dangerous reaches called *SOB rapid*.

In light of what I observed in this first trip on a large river, the pools and riffles, usually spaced at distances of 5 to 7 river widths, are regularly spaced in the canyon only in limited reaches and only in the lower end of Lodore Canyon. In the Echo and Rainbow parks the channel is sandy with many sand bars, dunes on the bed, and no deep reaches. The main canyon is so rocky or full of large rocks that the river cannot move. In the whole canyon, the maximum depth was 42 feet in a narrow section and just below the constriction entrance. The average depth we experienced was 4 to 5 feet.

Pools and riffles, when they occurred, had depths, respectively, of 15 to 18 feet and 2 to 3 feet. Velocity of the river averaged about 3 to 4 feet per second but in the rapids reached 10 to 12 feet per second (7 to 8 miles per hour). In the parks, the boat often fought a headwind that kept the speed to only 1 ft/sec over considerable distances. Pools tend to be deep in the upper end, shallowing downstream, plunging into the deepest part at quite a steep angle. The steepest portion of the whole distance was about 35 feet per mile at SOB rapid in Split Mountain.

Some characteristics of a river in a narrow gorge were worthy of note because I had not been sure that the things I had been experiencing in wide valleys would apply to a confined channel flowing between rock walls.

Cataract Canyon of the Colorado

My first river trip, down the Green River from Brown's Park through Lodore Canyon past Split Mountain to Vernalhad, had been an eye-opening experience for me. It was accomplished before the construction of any big dams on the Green and it provided a picture of a natural river that within a few years was to be foreclosed forever. Even though I could not have foreseen what engineering would be done on many western rivers, the fact that Glen Canyon Dam on the Colorado was nearing completion gave some warning of things to come. It was perhaps both a premonition of the shortness of time for continuing river health and also for the experience of a wild river that made me anxious to expand the experience to additional examples.

In 1964 I arranged again with Hatch to take a party of scientific people down the Colorado River through Cataract Canyon. For this trip I assembled some of our most dedicated research men, Emmett, Walter Langbein, and Herb Skibitzke. We were to meet at a place about 10 miles downstream of Moab. But before we gathered, I met Herb in Grand Junction and, after a talk with Charlie Hunt about his work on the origin of the Grand Canyon, we flew in the T-2 to Moab, Utah, but on the way made a big circle to the north.

On the way we went over Behind the Rocks, to Huntsville, up Waterpocket Fold, and around the north side of the Henry Mountains. The badlands at the foot of the Henrys are the largest, most gullied, and rilled topography I have ever seen. The hills called *Badlands* near the Black Hills of North Dakota are small and undistinguished compared with this large-scale, steep, deeply grooved hills. We ended by flying up the Colorado over Cataract Canyon.

I wanted to get a feel for the big river so I asked Langbein to drive Emmett and me well upstream where we launched the canoe.

The water seemed exceptionally smooth until we came to the first rapid, which gave us a little trouble due to waves that we seemed to ride over, making the canoe very tippy. In the tail of the next rapid we apparently went from the center of the fast water to the edge, where it was smooth but looked as if it were boiling up in a shear zone. The bow of the boat went over a large boil 15 feet across and very close to the edge of the fast current. It felt as if someone just pulled the left gunwale down because we went over like a round log. I was obviously surprised because it had looked so innocuous. We were in a long pool and I foolishly tried to swim toward the shore with the boat in tow, but in a moment I was tired and panting. Somewhat panicky, I calmed myself, grasped the stern painter where it was tied, and with the other hand pulled the life vest close to my chest just as we started into the next rapid. I brought my feet up partly encircling the stern, but the canoe rotated about its long axis as it also turned sideways to the flow. I watched Bill go under as the boat turned bottom up but he came up again. A paddle passed me as well as Bill's hat, but I still had my paddle in the hand that held the life vest. We passed a fisherman on the shore who called, "Do you need any help?" We could have used some but we were too busy to answer. I told Bill to just hold the bow rope and ride it out. After about half a mile we were swung into an eddy and my feet touched bottom. We got out cold and a little shaken. The depth sounder was broken.

We lined the boat around the next rapid that we felt was too big for us. The lessons: Boils, kolk, and shear zones are more difficult than waves. They can pull one side of the boat down, a phenomenon we had seen on smaller rivers. Do not try to right the boat or pull it by swimming. Be prepared to let it rotate and roll but hang onto it. Keep the life vest close to your chest so it can keep your head above water.

Where we launched the rafts the river was about 300 feet wide, averaged 30 feet deep, at velocity of 4 feet per second carrying 35,000 cubic feet per second. Charlie Hunt told me he continues to amass evidence that the canyon existed in the mid-Tertiary time, which means that in 10 million years or so the river has moved laterally only hundreds of feet while it downcut a thousand feet. At the same time tributaries of only 3 or 4 square miles of drainage area are cut as deeply into the bedrock as is the main river, for they are graded to the master stream. Thus even a small amount of water could deepen a tributary as rapidly as the Colorado River could cut downward.

I devoted some attention to this unexpected relation and it appears that the main river has the power to carry away all the debris supplied to it, but the tributaries cannot do so. I note that the main canyon has many vertical cliffs and only a limited number of talus slopes, whereas the tributaries are characterized by debris slopes and talus. The limited power of the small streams will downcut, but will not remove the debris that continues to accumulate.

In the first couple of miles after launching, the depth sounder was working. In a wide stretch of channel I measured depth with an oar and found it to be 8–9 feet. But shortly thereafter the instrument failed.

All the first day we churned along in still water at a uniform speed for there were no rapids, only a few riffles at this high stage of the river, and these hardly put the instrument to a real test, so the failure was a disappointment. There was one possibility. My colleague Herb Skibitzke was a mathematician and skilled electronics man. I asked Herb if he could look inside the bulky instrument and see if it was something that could be fixed, but we had no tools. At noon we heard our plane, the T2, fly over and Herb reached in his pocket and pulled out the smallest radio I ever saw. It was half the size of a package of cigarettes, and had a short wire sticking out of it. He put this thing to his face and spoke into it. One of our airplanes from Herb's lab was following us. He apparently described what he wanted and ended by saying, "Drop it at 5 p.m. this afternoon at the mouth of this Canyon."

A couple of hours later, 5 o'clock, I heard the drone of an aircraft and in a moment I looked up the narrow canyon and saw a plane in between the walls of the cliffs, coming at us at about 500 feet. Out of the plane dropped a parachute, pretty as could be, with something hanging from it. I hurried to the place the chute landed and picked up a small green steel box attached to the cords of the white chute, that, upon inspection, was not a standard parachute but a sheet from a nearby motel. Herb took his tools out of the box and spent a couple of hours at it. Finally he took me aside and said,

"This whole operation is not up to our standards. Let me get you our own boats, our own gear, and we will do all this the right way. You get us a boatman, I'll do the rest."

"Let's go," I said.

I had a talk with Ted Hatch and told him I was much impressed with the elder boatman, Smuss Allen, and suggested that because he was used only when Hatch had a river trip, I would hire him in those times he was free. I assured Hatch that we would never interfere with his commercial operation. Some months later Smuss Allen was on our payroll and found it was easier to work for scientists than to wrangle tourists on river trips. Smuss remained with us full time until he retired.

That evening Bob Myrick, Bill Emmett, and I walked up the canyon of this tributary. It was only 400 feet wide carved in red sandstone, probably of the Morrison formation, its walls hundreds of feet high and vertical. The valley floor was of a tan-color silt and graded to the silt banks of a bankfull stage of the Colorado River. In the lower 200 yards the stream had cut a gully some 3 feet wide at the base and 6 feet at the top, the bottom of which entered the main river at about the stage of present flow. This inner gully channel was discontinuous for it alternated with uncut reaches of the valley floor. Farther up the tributary the alluvium feathered out and the valley floor was bedrock. I noted that this seemed to be typical of many of the small tributaries.

The valley fill of alluvium met the vertical walls that enclosed the valley at an angle of 90 degrees, making it appear the valley is aggrading or still building up, though the trenching by a central gully indicated that aggradation had ceased. The most interesting geomorphic feature was the occurrence at several places along the valley wall of a form carved in the bedrock that resembles a fan for it was a round buttress, the sides of which sloped upward at an angle of about 30 degrees to disappear at a definite height above the valley floor. It resembled the talus form seen on knobs sticking out from a cliff wall. Those seen earlier were covered with talus in contrast to the re-entrants that produced a fan. These fan-shaped forms of bedrock must have been protected from erosion by a debris cover that lasted long enough for the cliff to weather back leaving the debris covered talus slope sticking out from the receding cliff.

At some later stage the protective cover of talus was removed, exposing the bedrock form it had protected.

The principal geomorphic problem presented by this canyon country is the vertical downcutting with practically no lateral migration of the channel. This is exemplified in many places but the most spectacular is the great horseshoe downstream a few miles. If the canyon existed during part of mid-Tertiary time, as Charlie Hunt argues, then the river had more than ten million years to downcut but moved laterally or sideways only a few hundred feet. I have commented elsewhere how the debris-rich tributaries compare with the more naked big canyon walls, but that is not visibly the reason for the lack of lateral cutting. The Horseshoe Bend in the Green River is a nearly 360-degree bend and the neck at the place where the curve begins and ends is only a few hundred feet apart, but the curve surrounds a ridge about 1,000 feet high and only a few feet across at the top. As we went around the long curve it was difficult to find just where we are. In this reach the cliffs became higher and the canyon more narrow. Soon we came to the beginning of the big rapids of which there are three. The river is only 100 feet wide and in some of the rapids section only 100 feet. Later we camped on the left bank at the mouth of an entering canyon that is small and narrow. We got started in the morning in high spirits and quickly approached the first rapid. In the first one everyone was shouting and laughing as we got drenched and very cold for the wind was strong and the sun only intermittent. To inspect the prospects, the boatman walked down the right bank over a rocky flat fan. Just before we hit the Big Drop we landed and everybody walked the left bank to see. One rapid is fixed in my mind as dominated by three big rocks, all more or less submerged.

The largest on the right had a hole below it 15 feet deep and with a tremendous roller on the downstream side that poured water upstream into the hole. It was obviously imperative to keep far enough from the side slopes of the hole to keep from being sucked into it by water falling laterally into the hole. But the rock near the center also was followed by a big roller that prevented one from steering wide of the first. Yet it was necessary to get left of the main convergence waves, which were immense, to be in a position to enter the next rapid. Smuss' boat led and I saw him skim the rock on the left but then gun the motor to miss the big waves. Our boat followed closely but a hair to the right of Smuss' position, and the big wave clobbered us. The bow hit the wave head-on, the boat bent double, and Bill, Walter, and the food box were knocked backward. The bow compartment was filled instantaneously to the gunwale as the whole top of the wave came in the boat. An oar from our pontoon was yanked off the oarlock post and went overboard. At this moment the motor was swamped and stalled. Everyone with an oar grabbed it, but meanwhile we had canted crosswise to the flow and we were going into the next rapid at 45° with the bow directed at the right bank. Ted yelled "back water!" and Herb and I pushed to beat hell. By this time we were in the center of the second rapid but three oars were beginning to be effective for the boat had reoriented and faced downstream, but dangerously close to another big rock. We now pulled valiantly and on my next upward glance we were in a fast but relatively smooth reach. By some fast oar-work we swung against the left bank, Bill jumped out and rapidly secured the painter on a big rock.

On a subsequent trip I was able to measure the depth of the water progressively along the channel. Never previously had such a profile been obtained through one of the roughest rapids of the Grand Canyon system. It revealed numerous violent changes in depth through the few miles near the Big Drop and the shallowness at the lip as the water plunges down into the hole below the main fall.

I had earlier asked Smuss whether this trip was routine in any way. He had negotiated Cataract Canyon more times than anyone alive. "No, it is somewhat different every run. Each time you go through the Big Drop you handle it as best you can, and come out at the bottom licking your wounds." We bailed out the boat, righted the food chest, recovered the lost oar, replaced the hanging floorboards. I got out my cruiser-coat and put it over my wet shirt for warmth. For the remainder of the rapids that day, our boat went through in tight-lipped silence. Camped that night at Gypsum Canyon, a wide, very interesting tributary from the east. It was an early camp so there was time to hike.

The canyon is dominated by a great debris flow that was later dissected by the main channel by small side notches. I heard an airplane approaching and, looking at the strong crosswind, Herb said into his microphone, "Do not drop!" but apparently Ray didn't hear the command and a bright-red parachute blossomed below the plane. As Ray pulled up for altitude, the chute gradually was blown toward the north canyon wall, but I could see it had just missed the cliff though its resting place was hidden. Bob and Bill started the long hot climb. Walter and I walked up the slope of one of the tributary debris cones, looking at the rillenstein-solution-faceted limestone on which rainwater had dissolved little channels that crossed the hard rock in an arabesque of patterns. The boys reached the chute and, hot but triumphant, caressed the cold beer down to camp. Next day we went through the Gypsum rapids, then the ones at Dark Canyon where the fall is sharp and steep, but less dangerous than Big Drop, in part because there is smooth water both above and below the rapid.

We were at Hite by noon the next day—the reservoir had been rising and had now reached the mouth of the Dirty Devil. All hands were depressed as we thought of these canyon walls sucked under the dirty rising pond of driftwood, flotsam, jetsam, and silt.

Ray was at the floodplain airstrip, that stood only 2 feet above the rising water. No doubt this is the last plane ever to land there. We flew out in two shifts. Bill, Bob, and I started on the road north to Wyoming, the taste of Cataract Canyon still on our lips, and the destruction of Glen Canyon heavy on our hearts. Back in the office while writing up the reports on these excursions, I began to see the possibilities, not only for reconnaissance and mapping, but for field support of expeditions.

Herb had been a pilot in the Navy and during the war he was flying heavy aircraft from Brazil to Africa and other delivery routes when planes from the United States were needed in the European theatre. He became acquainted not only with the internal operations involving military aircraft but also with many officers in that part of the service. Herb was working on the problems of obtaining boats, preparing for aerial photography, and transportation of equipment. He now had the possibility of mounting a full-scale scientific trip in the Colorado River system, including the Grand Canyon.

As part of these preparations Herb's research laboratory prepared its darkroom with a photographic capability to handle aerial

LBL and one of the US Geological Survey planes used in his pioneering field studies.

photo negatives of the usual 9" x 9" size. The Cessna 182 was equipped with a modern camera holder from the Army that would take photographs constantly horizontal, independent of the attitude of the airplane. Our De Havilland Beaver was to be the main logistical platform because it could carry a very large load and take off from unusually rough places.

This trip was the first successful measurement of the depths of water through a long reach of channel in a canyon, and, in particular, the depth characteristics of one of the biggest rapids in the Grand Canyon System.

Russia

There was a time when the administrators in the United States as well as scientists and academicians recognized the cultural accomplishment of the United Nations by its subsidiary UNESCO (the United Nations Economic, Scientific and Cultural Organization). Not only did UNESCO work to preserve ancient monuments, such as the Egyptian statues along the Nile, it organized research and even created research laboratories, and greatly contributed to the task of bringing representatives of various nations together to discuss and attack common problems.

UNESCO had several branches dealing with different subjects. The branches that concerned and interested me were those relating to water, climate, and hydrology. Members of the US Geological Survey made our contribution to these efforts by attending international meetings and contributing scientific analysis and papers. We even made a detailed geologic, geographic map of Saudi Arabia and so contributed to the finding, analyzing, and developing the ground water resources of that arid country.

Major efforts to concentrate the work of many countries are illustrated by the Hydrologic Decade, patterned after the shorter effort in the Geophysical year. In this case, the idea and its development came from my associate, Associate Chief Hydrologist Raymond Nace.

The unit to which I was invited was the Committee on Arid Zone Research, a general subject on which hydrologists of the US Geological Survey had given considerable attention. One of the real accomplishments of this committee was the establishment of several research laboratories, one of which was that near Beersheba, Israel, a location notably arid but in a geographic setting where water problems are permanent.

The work of the Arid Zone Committee was actively pursued during the Cold War when the chill of competition was only occasionally softened and nearly always by scientists, not administrators. This committee held conferences or meetings in many countries, but the one in Russia was most interesting and is the subject of the present discussion.

In a previous meeting of a similar kind, held in Madrid, the Russian delegation gave several papers in Russian, very few in English, and though the translation facilities were good, there was nothing that I could understand scientifically. The eminent biologist Dr. Evenari of Israel, who had long experience with scientists in different countries, said to me, "You know, the Russians do not mean for you to understand." So my colleague, Ray Nace, and I were wondering how we would be received in their country.

In London airport one could easily pick out all the persons who were waiting to board the plane for Moscow. Certainly, the

Russians were easy to distinguish, and the rest of the passengers were from Ghana or some other part of Africa, a few from India, and some from southeast Russia who were either Tibetan, Chinese, or looked like Navajo Indians. Ray and I were the only passengers in first class, but that word did not apply here because the plane must have been converted from a military craft. The inside was finished in dark enamel resembling wood veneer but looked like cheap plastic. Indeed, part of the finish was in gray plastic and brass. The brass formed heavy, plain, capitals on columns that were merely in bas-relief. The seats were upholstered in a heavy stuffed material that looked like cheap furniture. But the contrast with British Airways that brought us from New York was most striking in the stewardesses. The British always were young, attractive, and neat; those on Aeroflot were fat, sullen, and in soiled uniforms. What food they served was in cardboard boxes and not very appetizing at best. The plane was pressurized to 3,000 meters, which means there was much less oxygen than in American crafts. I was worried about ears or sinus but it was not bad for people who are young or not old. The pilots seem to climb to high elevation very fast, much faster than the cabin temperature adjusts, so there are changes from hot to cold. The flight did not seem to take very long.

Below the altocumulus a wet green landscape was revealed, with thick spruce, birch, and wet fields, a scene that appeared too cold and arid for growing good crops. The plane banked steeply and came in on the only runway, long but rough. There was a close inspection in the airport, they took our passports away, and appeared to make a detailed list of everything they could see, a ring, watch, and what else. The airport is some miles from Moscow, traffic moves very fast but was not heavy. Pedestrians have to run to cross the street and drivers do not seem to worry much about them.

The hotel was very large, run by the state tourist bureau, which also functions as the intelligence bureau—as became clear. The hotel was full and there seemed to be a lot of groups but very few Americans. It was interesting that every room in the hotel that I saw was decorated with oil paintings, not prints, and they were good, though there was no modern art. It is a massive building reflecting the bulky and uninteresting architecture of the Cold War Russian program, and within the structure, there are frequent failures of water or heating or even structural failure. I had the feeling that because I was the chairman of the UNESCO delegation, I was being watched, even spied upon. The feeling was reinforced when, in a few days, we traveled to Tashkent, Uzbekistan.

I learned from Michel Batisse, our representative of UNESCO, that we were to go in the morning to the offices of the Meteorological and Hydrological Institute. Michel, Ray, and I arrived at 10:30 and were given, in lecture style, a general briefing on how the international proposal was developing, an attempt to coordinate all observation stations in the hemisphere so that all countries will have the worldwide data for forecasting and record keeping. The Russians felt that this was too broad a mandate and that each country should coordinate its observatory program in its own way. The attitude was typical of the Russian point of view, even on matters of scientific observation. We have maps of the locations of river measuring stations in the United States, so it was logical for one to ask to see a map of similar observation points in the Soviet Union. I was told this was a state secret and not available. I told the officers we had about 18,000 observation points in our river-gaging network, including stations that had been terminated. I asked how many such stations were in the U.S.S.R. This too was a state secret. It was not a very uplifting morning but left one with a keen impression of what secrecy is doing to science.

The next day we were taken to meet the minister of Cultural Affairs. He spent a couple of hours telling us how the Ukraine had been devastated by the Nazis and rebuilt after the war. He wanted to know our impression and urged us to say whatever we wished to see in the city and its environs. We expressed interest in visiting the outer fringes of Moscow that had not been changed by modern buildings. Not unexpectedly, our request was ignored.

Later we went to the museum within the walls of the Kremlin where the exhibits are principally concerned with the Czarist period. This was absolutely fascinating, for the grandeur of the buildings is matched by the beauty of the objects exhibited. The famous decorated eggs were displayed admirably. The collection of the armor from the Tartar period (1200–1500 AD) I found most interesting. I was fascinated to observe an old woman, apparently from the lower class, looking at the golden extravagance of the days of the Czar.

She was as impressed as I, and apparently appreciated the display as much as anyone.

The same day we explored the multiple cathedrals within the Kremlin walls, dating from about the 16th century. Some of the icons, of which there were thousands, dated from the time of the infamous Theodora, spouse of the emperor Justinian in 550 AD. We returned to the center of Moscow by riverboat and I still remember the captain of the vessel, a beautiful young woman who was the only person with a smile that I saw in the city. There seemed to be a general lethargy in the society, both cynical and tired.

Friendly faces were nearly absent. One small incident provides a glimpse of what might be an undercurrent of some promise. Ray Nace and I were walking back to the large hotel after a stroll in the town center. A black car drew to the curb beside us. All the cars in Moscow seem to be black in color and of the same design, built by the same company. The driver of this car, whom we thought to be a taxi driver, motioned us to get in and seemed to understand our destination. Without hesitation he drove to the hotel using some streets I would not have known about. At the door of the hotel we stepped out and I took out my wallet to pay, but the driver just waved his hand and drove off. Contemplating this we decided there were many decent people in the country that do not hate America.

On one of our walking excursions we took the subway to our destination and admired every aspect of the train stations, the trains, and the cars in which we rode. Inside the station, the architecture is lovely, in some places with graceful archways, and tastefully decorated, some with tile, and some with paint, in bright but subtle colors and designs. The stations had beautiful and elaborate chandeliers, never seen in an American public place like a subway. We were truly impressed.

I had dinner this last night in Moscow with Dr. Vanderburg, committee member from the Netherlands, and a UNESCO journalist at the Hotel Nacional, the best and only good meal I had since we arrived.

The next day we flew first to Sverdlovsk, about 800 miles west of Moscow, a large industrial city. That leg was similar in vegetation to the area around the capital as far as the cloud cover allowed me to see. After a short stay the plane proceeded south to Tashkent, located in the far-east corner of Uzbekistan. The city is dry and dusty but the people are handsome and very different from Russians in the Moscow region. The Uzbek men nearly all wear a mustache, and wear on the head a round cap called a *Túbitál* usually embroidered around the band like the Jewish cap but extended down all around the head like a cup.

We were taken by bus to the one hotel located at one end of a rectangular park at the other end of which is an imposing building, the main opera house. In the center of the park is a large fountain throwing tongues of water 30 feet in the air. I was assigned a hotel room on the second floor, the window of which was centered on the park so I looked out directly opposite the center of the opera house. There is a keeper of rooms on each floor and she has the keys to the room. Therefore there is always someone who keeps an eye on the guests and probably is also distantly connected with the KGB secret service.

That first evening I went to dinner with Harold Thomas, a professor at Harvard in engineering and statistics whom I had known briefly when I was a student in 1937. Thomas had an Uzbek friend named Rámir whom I befriended and who translated for me. We went to a restaurant chosen by Rámir, where the customers were all young people who came to drink and dance as a US-type of band played very danceable music. Rámir ordered a carafe of vodka, holding about a full pint, and consumed it all in one evening. I was very interested in the dancing, which was the American foxtrot, very little twirling or flinging, but more constrained and graceful.

My attention was drawn to one very beautiful young woman who was clearly an accomplished dancer. I saw various men cutting in on a couple and I asked Rámir if this was acceptable and common, and he said it was standard procedure. I decided to cut in and dance with that girl even if I could not speak a word to her. She was gracious and I immediately found out that she was the best dancer I have ever met, graceful and smooth as well as beautiful. Rámir danced or talked with her and learned that her name is Rima, that she is a graduate student in physics at the local university, and that she comes from Dushanbe, capital of Tadzhikistan, which is about 200 miles south of Tashkent. Rámir told her something about the meeting we were attending.

Even at the end of the first day it was apparent that I was being watched, perhaps tailed, by the police or more likely by the KGB. In my hotel room, the telephone rang and when I answered it, there was a silence, then the caller hung up. On other occasions, a woman's voice would ask, "Professor Leopold, are you going to the meeting? I will send a private car to take you there." They had no intention of letting us be by ourselves any place or any time. The way the woman key keeper on my floor acted toward me strengthened that impression. But at one time she indicated that she had prostitutes available on the same floor.

The following day the conference met in the Soviet headquarters in a round auditorium where there were good audio facilities but the seats were like Greek marble benches without backs and uncomfortable as hell. There we met Dr. Korda, who held a high position in the Russian science hierarchy and whom I had met previously in Washington. I got the impression his standing in Russia was more of the political aspect of science rather than of science. We also got to know Nelli, the principal translator for the Committee meeting, a sweet girl who spoke quite good English and was quick and apparently accurate. I, as chairman of the American delegation, and a few others were seated at the Praesidium of the auditorium, which means the front table behind the rostrum. Though there was simultaneous translation into English and French, the Russian speakers were very long-winded and tedious and the translators were often hard to follow, so the meetings were frankly boring. At the morning break, those people at the front table were invited to a coffee, with real fruit, good peaches and tomatoes but poor grapes, apples, and plums. They also had excellent chocolate. I found that the best food in Tashkent was Uzbek bread, baked rather like a bagel, round but with a hole in it like a doughnut, with a wonderful texture resembling a French bread.

Usually the only meal I ate at the hotel was breakfast. It took some time for the kitchen staff to understand that all I want is an egg and Uzbek bread. When they finally got the idea, my breakfast would appear with a tiny American flag sticking up out of the egg.

The Russian government had more or less eliminated the practice of Muslim religion in the country, but I learned that here in Tashkent was one of the few remaining mosques anywhere. I asked

Nellie if she could arrange for my committee to see the mosque and possibly meet the mullahs. This was arranged and we were taken by bus across the city to a very small white building in a garden setting, and were invited to look in the building. It was very simple and had no decoration, no tiles, but elegant. At the entrance I could see a crowd of people kneeling on the floor facing the interior. At the pillared entrance there was a multitude of footwear because a service was going on. The shoes, slippers, sandals, scuffs of all kinds indicated the participants were far from wealthy but the whole scene was one of relaxation, of serious contemplation, and so was to me quite moving. I took off my shoes and stepped in the portico just far enough to see the small crowd of men kneeling facing the same way and to hear a low buzz of prayer that was obviously sincere and respectful.

We backed away from the service and wandered through a small but delightful garden filled with flowering plants carefully tended and tastefully arranged. A subsidiary building was at the far end of the garden and in its bright and shady interior I found that the hosts had put out on long tables old books and manuscripts from their collection for us to inspect. To me this was especially meaningful for, as a hand bookbinder, I found the various types, conditions, and forms of binding and decoration of great interest. Probably the most unusual object on display was a copy of the Koran that a small explanatory note said dated from about 680 AD. I observed one of our committee rustle through that volume as one would a new paperback and I fear I spoke sharply to him.

Soon we were called to a room in an adjacent building where a delicious repast awaited us. The room was expansive with a high ceiling, decorated in part with attractive columns, with the floor covered with exquisite Persian carpets, of the most intricate design and beautiful color. So also was a long table along the length of the room, covered with delectable bowls of fruit of many kinds, and presented in the most beautiful way. After a pause to let the guests eat, one of the mullahs, with a long white beard in a white turban and flowing gown, spoke to the group apparently in Arabic. I asked Nelli if she could tell us what he said, but after a few words she began to cry because the translation was beyond her. Someone else intervened and translated the Arabic into Uzbek, and after recovering, Nelli translated it into English from the Uzbek. The words were a graceful expression of welcome and wishes for success in our visit to Tashkent. I then spoke briefly of the beauty of the country and the delicious fruit, and I thanked the mosque for the hospitality offered.

Returning to the hotel there was a chance to swim in the nearby lake, where most of the committee gathered along with Nelli and Paris (my name for the wife of one of the UNESCO staff). This was pleasant even though the lake water was dirty. There followed a reception at the Academy of Sciences, a branch of the Russian National Academy, with lots of vodka and many new people to meet. A few of us were to go to the operetta. Thomas and I took a taxi because we were somewhat late for the opening. The operetta was presented in an open-air stage, the audience sitting on benches in the cool air as the sun went down. The stage appeared well equipped with overhead space above from which curtains could descend and scenery could be lifted or dropped. As the French member of our committee said to me, these people thirst for culture and knowledge. We had waited at the hotel thinking Nelli was going to interpret for us at the opera house but though we came in during the first act it was quite possible to follow the action. The audience was all local Uzbeks with the traditional *tùbitàkas*, men, boys, old people, everybody, all whispering to one another in a respectful buzz. They seemed to be following the action and anticipating the next scene. What was happening? Here comes the first Prince! He is going to get the girl in the end, don't worry. Also there's that bad vizier. He's a rogue. Here she is, this is the princess who was kidnapped. An old Uzbek lady next to me nudged me and smiled, with the usual lot of gold teeth. She whispered, "I'm sure; watch the first Prince, he will come out with the girl in the end. Don't you worry." The whole audience, which I watched as much as the play, gripped the back of the next row of seats and leaned forward.

The dialogue was in Uzbek, the music somewhat oriental, for there were native instruments rather similar to those we know. Some stringed ones were on sounding boards of strange design but beautifully made; among them percussion types, the strings being struck by special wooden hammers. Singing was with a kind of waver of the voice at the end of a phrase, similar to Greek songs. The characters in the play sang rather than spoke the dialog. The only dance in the play was like what one sees in India. The setting of the play appeared modern and props were lifted up by ropes, as in our theatres, but the stage setting was suggestive rather than realistic. There were apparently only two persons in the audience who knew both Uzbek and English, Nelli and another interpreter. The two could not agree on many details of the plot of the play, but we could follow it without knowing a word spoken.

In Russia there were many soil specialists who had made many important contributions to pedology. I heard there was a Soil Science Institute in Tashkent and not far from the hotel. I walked down there but nobody answered the door. Later I was told the Institute had closed. A couple of days later I was attending a scientific meeting when someone tapped me on the shoulder and said that the Soil Science Institute was here to see me. Surprised, I felt that someone should accompany me to such a meeting and I asked the committee member from Chile to go with me, for I knew he had an interest in soils.

We were taken to another room and I found it full of at least fifteen people. Trying to get started on the science aspect, after thanking them for coming, I described what happened in the United States in the last part of the 19th century when an epicycle of erosion began on the western part of the continent. Soils were eroding in all the valleys, massive gullies developed, irrigation diversions were isolated, and economic hardship followed. The cause apparently was overgrazing by stock, combined with a palpable change in climate. I then asked if Russia had experienced anything like that in their recent history. After a long pause, one person spoke slowly and with care. "We have in the Soviet Union many soil experts who work in every part of our country. We have many laboratories engaged in analyzing soil."

"Did you experience any change in the stability of your soils?" I asked.

"We have many experienced people who work on our problems," was the reply.

I persisted. "Was there erosion?"

"Many of our scientists work long days on our soil problems," he said. Another person spoke up: "We are all specialists in soil work and we solve many problems," she said. That was the tenor of this exchange. Later I was able to answer the question by my own observations.

There had been much talk in the technical sessions about the use of water for irrigated agriculture, which, as we know, consumes more than three quarters of all the water use in the world. The committee was taken to see the irrigated areas of cotton located north of Tashkent. There were miles and miles of cotton plants on the flat area bordering the Karakum desert, irrigated by water diverted from the Syr Darya. The two rivers that flow into the Aral Sea, the Syr Darya (Oxus) and Amu Darya, both rise in the high mountains of Tadzhikistan and both are progressively depleted for irrigation until, as we now know, there is no flow into the Aral Sea. That small sea once had a lively maritime culture with a good yield of fish, especially sturgeon. Irrigation has so depleted the rivers that fed it that rusting ships stand in a desert of sand. Our view of the fields of irrigated cotton told the story at a glance. The irrigation system was poorly built, badly deteriorated—even when we were there—and obviously inefficient. Furthermore, cotton was not a high-value crop. Large-scale irrigation could hardly be called a good use of scarce water in a desert.

I think back to a dinner party given for our committee in a raised open but roofed platform located in the middle of a cotton field. This round wooden structure seated about 30 persons. At the end of a day, we sat in the field with but little breeze and the mosquitoes were voracious as well as numerous. As at many parties in the Soviet Union, it all begins with a toast that, to my astonishment, started around the table, each person giving a toast with a shot-glass of vodka, but some of the speeches turned out to be lectures. After two shots of vodka I looked around for ways of disposing the drinks and saw that the floor of the platform was of boards set about an inch apart, so one glass after another went down the cracks to the cotton plants below. The natives seem to thrive on these free drinks and apparently downed every glass as it was refilled.

The last days were the most rewarding, both scientifically and culturally. The committee flew to Samarkand, on the way passing over the southern end of the Karakum, an arid landscape that includes some dunes of sand, not active in the present climate. On this trip I could see from the air that valleys and draws that in the arid southwest of the United States would be gullied or dissected were here smooth and uneroded. In the light of what I knew about climate, it was clear to me that there is no summer rainfall here and there are no thunderstorms with heavy bursts of rain. This to me answered the questions I had posed to the Institute of Soil Science.

There are so many things about the history of Samarkand that made it a real thrill to go there. Alexander the Great had led his troops north over the great mountains of the Hindu Kush as far as Samarkand. On that trip in 328 BC he had stormed the local strongholds of northern Iraq, one of which was where he met the girl with whom he fell in love and married, Roxane. She was the daughter of an Iranian baron Oxyartes. She bore him a son who was doomed to die because so many of the generals wanted to inherit the countries conquered by their leader, and on his death they certainly did not want his son to be in their way.

Just 1663 years later the same places visited by Alexander were the operational center of a race of Tartars, warlike horsemen from the far north, led by their chieftain, Timur or Tamerlane. This man, quite like Alexander in fact, conquered much of the known world but, amazingly, left little permanently to show for it. Alexander had left his mark in many ways, notably through cities he founded that have grown to the present day. Possibly because Timur came from nomads whose horses were their occupation and practically their home, they were not settled people, nor settling people. So to view the city that was Timur's home and to see his tomb was a cultural high in one's life.

Samarkand was known to the ancients as a land close to Paradise, rolling hills covered with meadows, cultivated fields fed by the Syr Darya, close to the high mountains that fed rushing streams. The river was the traditional dividing line between Bactriana (and Iran) on the south and Scythia (now Russia), a land of cattle breeding and horse raising. Everywhere in present Samarkand are abandoned buildings of sunbaked bricks, a type of construction unfortunately subject to destruction by earthquakes, the results of which are apparent. The beautiful domes covered with blue tile that shine in the warm sunlight are the most striking things about the city. There are arches of medieval forms as well as different domes over large buildings. Some must have been built by Timur in the 14th century, especially those covered with blue tile. Some domes are nearly hemispherical, others are shaped like the stem end of a pear, grooved in vertical ribs converging at the top. The most common shape seen in Moscow is the bulbous dome, swelling out from the base, and tapered to a slender point. I read that Timur saw this shape in Damascus while he was destroying that city and had his architects copy it. Many of the churches within the walls of the Kremlin are of this type and most are covered with gold. St. Basil Cathedral on Red Square, probably the most famous building in Moscow, has decorated towers of this shape. In Samarkand there are remnants of high blue-domed structures, now destroyed, that have walls covered with white tiles on which are the exquisite designs of Arabic calligraphy. The tomb of Timur is a simple but elegant structure that gives the impression of an open, airy colonnaded vault. Timur was known to be disabled; some time in the recent past the sepulcher was opened in which were found the remains of a man whose legs were indeed of unequal lengths. At the time of our visit to Samarkand, archeologists had begun excavation of the astronomical observatory of Ulugh Beg, scientist as well as a ruler, the grandson of Timur. His grandmother was the great princess Sarai Khanum, and his father was Rukh Shah.

As explained by Harold Lamb, Timur, on his deathbed, had designated his son Pir Muhammad to be his successor, but he was leading the troops to India. The ablest son, Shah Rukh, was busy as the governor of Khorassan. The least trustworthy, Khalil, son of Miran Shah, was in Samarkand, and in the absence of the others, was declared emir. When Ulugh returned with the body of Timur, the gates of Samarkand were closed and he was defeated trying to retake the city. Khalil married the courtesan Shadi Mulkh, dissipated the city's wealth, and gave away its treasures. After a period of civil war, Shah Rukh finally returned, captured the city, and gave it to his son Ulugh, and between them they controlled the core of the empire from India to Mesopotamia. Even though the city was in ruins, these two men rebuilt the region and made it into an island of refuge, where poets, philosophers, artists, and intellectuals flourished.

Of all the children and descendants of Timur, these two were the only ones interested in cultural matters, not territorial expansion and conquest. Their interests were continued by their great grandsons, who founded the Moghul dynasty in India, now famous for its artistic treasures. The excavation of the observatory that we saw consisted of a trench, oriented due north, in which steps of marble were set in a quarter-circular arc on which the observer could sit watching the stars go by his line of sight, enabling him to measure the angle above the horizontal at which they occurred. Ulugh compiled an atlas of stars in the year 1540. It was the second such catalog, following that of Ptolemy, who in 141 AD, compiled the first star map.

The departure from Tashkent was a trial of nerves and frustration. After a long wait at the airport with no explanation for the delay, we finally were boarded along with a few chickens of an Uzbek woman with a reed basket of clothes and food. Once aboard the aircraft that had been sitting in the hot sun for hours, we were seated in the very warm cabin and just waited for something to happen—about two hours of it—perspiring and getting more anxious for at least an explanation if not for a little fresh air. Finally, the motor rumbled, the plane took off at a very steep angle into really cold air at 30,000 feet, and all the overheated passengers were literally freezing. I had the beginning of a cold and was truly miserable for many hours in this cold military aircraft in which there seemed to be no concern for the comfort or well-being of a full load of passengers. At Moscow airport, I asked Ray to look out for the luggage while I tried to find out where our passports were and how one checked in. The airport seemed to be a rabbit warren of halls with no signs. I bumped into the captain of a British Airways plane who told me he would be departing shortly. I begged him to wait for us. Finally, I found a large, blousy woman behind a card table in a hall who seemed to be one who handled tickets. When I asked her how I could trade my ticket for one on British Airways she acted just like those in a James Bond novel, and she looked the part, too. Finally, after some bickering, I took out a US government transport request and indicated I was buying new tickets. Finally, in full ill humor, she gave me a pass to the English plane, and I rushed to the exit doors where Ray had been long waiting. At the foot of the movable steps up to the plane, three security people gave us a final check, for what I never knew, but we mounted the steps and collapsed. Never was I more glad to be with people I could trust.

Ancient Sources and Developments

Groundwater is available in a large number of field conditions and in alluvial valleys nearly always. It also is a surprisingly large percentage of the total available fresh water on earth. Before the advent of the centrifugal pump, the resource was seldom stressed from human use, and shallow wells dug by hand provided an important part of all water use. In castles and other confined places, the usual source of water was a well in the courtyard from which water was taken in buckets, often by use of a windlass or other pulley arrangement, lifting water directly from the local water table. One often sees an open-roofed structure built over a well with a pulley hanging from the small roof and a rope passed through the pulley ending with a bucket in which to lift the water up.

I have tried to find the location of such a well in every castle, manor house, or village I have visited. Ancient castles, monasteries, or old building complexes nearly always have a well in a prominent place, many of which have water at the bottom even today.

Many such wells are lined with rock work to keep the walls from slumping and a few are dug into rock, tapping the water that will drain from cracks, joints, or other openings. Most had the entrance at the top surrounded by a low wall and often roofed over to prevent rain or objects from falling into the hole. The abandoned ones usually now have a grating to protect visitors from being harmed if they fail to realize that it is an open hole.

A typical one is found in the courtyard of the ancient ruin of the Monastery of the Cordeliers near St. Emilion, France. An unusual example is the Scottish castle Visimul located in the salt water lake Castlebay for protection. It is one of the oldest, having been built in 1039 but somewhat altered later. It has two fresh-water wells in the court. An interesting example is in the castle of Peniscola on the shore of the Mediterranean in western Spain. After years of unsuccessful entreaty, Pope Benedict XIII was finally forced from his fortress at Avignon and banished to Peniscola, where he was a virtual prisoner for the remainder of his life. This castle stands high above the shore and its principal terrace or courtyard is at least 75 feet above the water, yet the single well is in that courtyard. The well must have been drilled into bedrock to draw on the local water table, more or less uncontaminated by the Mediterranean.

Wells in alluvial valleys are often constructed for irrigation and these nearly always from a wheel that lifts an endless belt on which a series of buckets are attached. The buckets dip water out of the well, are brought to the surface, and tipped into a channel to convey it to a nearby plot. In India, these are usually powered by an ox, which walks around the well. This same principle is employed by one near Cartagena in southern Spain powered by a very large windmill built with huge vanes like those familiar to visitors to Holland. Another I saw in Portugal is a simple structure powered by a horse walking in a circle beside the well, not around it.

One of the problems in ancient structures, such as castles, was potential contamination by nearby water bodies, such as moats. Nearly all castles in England are surrounded by a moat that was filled with water, though many are presently dry. The source of water for a moat is often difficult to determine because many changes have occurred in the landscape during the centuries since their use. Usually a nearby stream was tapped, and a few were built where the water table was high enough to fill the moat. But disposal of waste in such closed structures must have been a problem. No doubt the knight or owner was provided with chamber pots by servants who dumped their contents into the moat. Others in the building used the open-air openings in the roof or high point constructed to overhang the moat as a privy.

Cities, however, were sometimes provided with a continuous supply of water that provided continual flushing of public latrines constructed with forethought for cleanliness and comfort. Long before the Romans had constructed the aqueducts carrying water for miles from the source, the Macedonians and others had sophisticated systems of supply and disposal.

I visited the ruins of Philippi, a major city built about 340 BC during the period of Philip II, father of Alexander the Great. In the ruins is a well-designed covered canal lined with rock that provided a series of stone toilet seats continually flushed toward the nearby sea. There was no obvious source for water, but on close inspection I found remains of a long channel that led around a nearby hill that tapped a spring or perhaps at that time a small stream about half mile from the town.

There were other landscape features besides wells that provided water at least for human and stock use though not for agriculture. These were usually small and often temporary: accumulation of water in ponds having a variety of names. In the southwestern United States they were natural rock tanks called *tinajas*, and pools where water spread out over adobe flats called *charcos*. A small earth dam forming a pool that held water is called a *tank* or *represo*.

The Indians of the southwestern United States made extensive use of tinajas and represos. At a Hohokam ruin called *Long House* in northern Arizona, there is a small dam, long ago breached, that formed behind it a tank that doubtless was a major water source for that community. North of Phoenix, Arizona, is the deep canyon of Aqua Fria bounded in some reaches by steep scarps of basalt, a hard rock of once molten lava. On top of one of these cliffs is a remote ruin presently reached only by helicopter. This village of ancient Indians was built near a small natural rock basin in the rock, a tinaja. But the rainfall is sporadic and at times the local women walked a narrow trail now mostly obliterated, to fill an olla with water of the river and then traverse the 1000 feet up the slope to reach their homes.

An interesting variation I saw in the Negev desert in Israel was a small water reservoir cut out of solid rock next to a dry wash. It was built by the Nabateans about 2,000 years ago. The tank had a hole on the side near the top that allowed floodwater in the channel to enter during a desert flood. The hole for refilling served as entrance during construction. On top of the cistern was a hole about one foot in diameter permitting a small bucket to be let down on a rope. It must have been widely used for a long time because the edges of the top hole are deeply grooved by the ropes used to bring up the water. The design had the great advantage of collecting only water from the top of a flood so there was practically no sediment accumulating in the cistern. Its age was indicated by the fact that in the cavern was a picture of a bull's head carved on a wall, a national symbol of the Nabatean tribe.

Past changes in conditions can often be estimated by the physical forms seen in local valleys, on hill slopes, and in banks of channels when they are high, as in an arroyo. The sequence of cutting and filling of alluvial valleys has been extensively studied, especially in semi-arid conditions, where the lack of vegetation allows good exposures of geologic materials.

Throughout much of the western United States, in the Mediterranean area, and in northern Mexico, alluvial terraces represent remnants of former flood plains of earlier stages. It is generally agreed that these changes in stream valleys are the result of climatic change, not necessarily of total annual rainfall but by the character of the rain during storms. High-intensity storms usually cause erosion of channels, whereas long continued gentle rain appears to result in surface erosion and leads to deposition of alluvial fill in valleys. This type of analysis was begun by Kirk Bryan and has been greatly extended first by his students and then by others. A recent summary of examples from the Mediterranean, Mexico, and the southwestern United States, was published by Leopold and Vita-Finzi.

The remarkable aspect of this history is that during the 10,000 years of the Holocene, periods of valley filling have alternated with periods of erosion not less than three times and in some places more. This established sequence should make it clear to present observers that changing climate has been the rule in recent centuries, though now greatly exacerbated by human activity.

A widespread phenomenon seen in the Mediterranean area is the presence of ancient agricultural terraces, now abandoned, constructed on steep valley slopes to provide enough flat area to grow some crops, even though such flats may be only a few yards wide and hundreds of yards long. Irrigation was impossible on such plots so success must have depended on the sporadic rain of the semiarid climate. These remnants of an ancient practice dominate the landscapes of much of southern Spain, southern Portugal, Greece, and most of the Greek islands in the Aegean Sea.

Why should men have gone to such effort to provide a pitifully small area to cultivate, to reach which he must have gone long distances, and on which the arable soil was generally shallow and not rapidly furnished with organic material to maintain tilth? A reasonable hypothesis is that the valleys where these terraces occur may at a still earlier time have had some flat land that was washed away by erosion. When there was no land in the valley bottom to cultivate, the local people had to build some land in order to survive. Some time later a climatic change caused alluviation of the valley by deposition, arable land reached with available water was created, and the terraces were abandoned.

I had a plan to visit as many ancient sites where water was intimately involved as possible. Of special interest is the water needs by great numbers of soldiers moving by foot over vast distances, a circumstance not faced for centuries before our time. The first of consequence in my search concerned the determination of Darius the Great, who ruled over the great kingdom of Persia in 500 BC, to conquer Greece, for which he constructed a fleet, and with a land army advanced from Persepolis across the Hellespont accompanied by his ships. In the fingers reaching south from Thrace, the three peninsulae of Chalcidice are mostly low-lying, except the most easterly one, the tip of which is dominated by the high mountain of Athos. As the large fleet advanced around this impressive feature it was caught in a great storm and destroyed as a fighting force. The expedition was abandoned.

Darius had two sons, the younger born of the king's second marriage to Atossa, the daughter of Cyrus. Atossa was a woman who had much power over her husband, and she insisted that her son, Xerxes, be named successor. When Xerxes became king he decided to pursue the failed project of his father. To do so, in 480 BC he raised an army probably more vast than any in history. Herodotus described how the king counted this horde, constructing a low wall surrounding a packed crowd of 10,000 men, and successively pressing units of that size in the enclosure and counting the total, which was 1,700,000 in the land army. The ships numbered 1207.

These numbers are probably exaggerated and modern scholars put the number of soldiers around 300,000; still a large crowd marching on foot and needing food and water.

Even before Xerxes crossed the Hellespont, he sent a large body of troops to Chalcidice with orders to cut a canal through the peninsula behind Mt. Athos so that the fleet would not be exposed to the storms that broke the back of the Darius expedition.

After crossing the Hellespont, the army of Xerxes proceeded through Thrace supplied with food by the cities through which it passed, bringing ruin to them. They reached the river Lisos, the modern Mista near Thasos, but the river was essentially dried up by the troops. At the city of Pistyros was a brackish lake about three and a half miles in circumference abounding in fish, but the baggage animals dried it up. At the river Strymon, still called by that name, there already was a bridge. The Thasians provided a dinner on behalf of their towns that cost four hundred talents of silver (nearly 23,000 pounds). Xerxes camped for some time near Therma (present Thessalonica) near the river Cheidoros (probably the Galikos) that was insufficient, "for drinking of the Army and failed in its stream."

While he was advancing through Thrace with his land army, the fleet sailed through the canal that he had ordered cut through Chalcidice. Herodotus described the cut having a top twice as wide as the channel itself "with such breadth that two triremes might sail through, propelled side by side." The width of the peninsula through which the cut was made was about 1.5 miles.

One of the reasons I went to Greece was to see what remained of this canal after 2,400 years. I found the cut, which I identified by geomorphologic evidence. At one place it was excavated for about half a mile through a foot slope of a mountain nose and the slope of the colluvium was interrupted by an unnatural break that outlines the original excavation. With no instruments I estimated the width to be about 150 feet at the top, and the bottom, though now rounded by sloughing, must have been about 50 wide. The bottom is now a tangle of brush. Also obvious was the fact that earth movement had lifted this central reach of the canal above present sea level. Scranton, Shaw, and Ibrahim state that the rise was 14 meters.

When Xenophom described the march of the ten thousand from Babylonia through modern Turkey to the Hellespont, that army either purchased food or commandeered it from local tribes and still had periods of insufficient food, but lack of water was not a problem to them.

The experience of Alexander the Great was different. Unbroken successes in the astounding march from Macedonia through Greece, Turkey, Palestine, Egypt, Arabia, Persia to Pakistan changed to catastrophe when he began to return. At the mouth of the Indus near present Karachi, Alexander planned to march northwest along the Persian Gulf accompanied by a fleet that was to provision the troops with both food and, if necessary, with water. The logistics were well planned and a similar combination of support to troops from ships had been universally successful.

Confidence in the plans for sustaining the horde is indicated by the fact that the 30,000 persons included "some women and children and a train of Levantine Traders" as described by Robin Lane Fox. Weather and an unexpected attack by Indian natives prevented the Admiral, Nearchus, from sailing on schedule. He was detained thirty-three days.

Alexander started into the Gedrosian desert believing that the grain ships would catch up with them, but suffering began quickly. The route was confined to a narrow strip of land between steep cliffs and the ocean. The food was depleted, the temperature about 95°F, even at night, and the walking troops encountered sand dunes difficult to cross.

"When camp was pitched it was kept as much as a mile and a half away from any watering place to stop men plunging in to satisfy their thirst. But many, still, would throw themselves in, dressed in full armor. They would drink like fish under water. Then they would swell and float up to the surface having breathed their last. They would foul the remaining water." (Robin Lane Fox)

They had been able to dig for water at shallow depth along the coast, but conditions of hunger and heat made him decide to turn inland toward the Gedrosian capital at Pura. The few passes through the harsh mountains were narrow and confined within high walls. They were forced to camp in the bed of the desert stream and at one place a flash flood occurred and many people drowned. Moreover the guides lost their way and, as Fox explained, "on the shore they faced famine, inland, thirst."

Modern armies have faced serious difficulties in fighting during terrible weather, as with Napoleon's attempt on Moscow, but the numbers of persons moving on foot for great distances is a circumstance probably not equaled since Greek and Roman times. Then water, more than any other necessity, has been the limiting factor.

6
Some Water Studies

In a career that involved not only many subjects but several professions, topics arose for investigation that were deliberately chosen by me and there were others on which I was asked to be engaged. Some are related to research and others to science administration. They all have in common the necessity of detailed observation, a practice that certainly is important in all the subjects in which I have ever been involved. In the ten years that I was in charge of the large organization of the Water Resources Division, much of my time was devoted to personnel, budget, and organizational matters. During that time, I oversaw the consolidation of administration into one in which a single person was in charge in each state, replacing a system in which the various disciplines—surface water, groundwater, and water quality—were each represented separately in a state. That was traumatic for the organization but far more serious was the feeling that my emphasis on research was inappropriate and underestimated the importance of the day-to-day data-collection effort.

The reason this was so disturbing was that no one in the organization had had any experience in research and the workers were not convinced that this was good for the Survey. Furthermore, I was introducing a form of geology into a group that had always been separated from the Geologic Division of the Survey. When I came in no one in the office had ever heard the term *geomorphology*. In order to explain the rationale for these changes I spent much time traveling from one to another among our 350 offices. We maintained cooperative arrangements in many foreign capitals and were concerned with many international water programs. These travels gave me experience in many water matters and led me to become acquainted with other scientists in the discipline.

To round out our roster, I sent selected men for advanced training at universities paid for in full by my office, offering the graduates an appointment on our staff. Some stayed only a short time with the Geological Survey, but those who left were in a position to further the hydrogeologic science. I experimented with inviting for a one-year period someone from a foreign country with all expenses paid to be on our staff.

I expanded the publication program and added to the physical equipment needed to carry on the work, with airplanes, boats, and instruments. In the long run the appointment of a small staff of well-trained people led to a great expansion of knowledge and interest in hydrology and geomorphology. The publications of the Survey became known as a scientific force in our field.

One of the most important of the ideas I introduced was the establishment of a network of benchmark gaging stations located in places where we thought they would never be subjected to land-use changes: parks, wilderness, or other protected areas. The idea was to see over time what changes if any took place under natural conditions. About fifty such stations were established representing different geographic, topographic, geologic, and vegetative conditions. Most of these are still in operation and their usefulness is now being recognized as climatic change occurs.

Through all this time, I was determined to see that my personal research would continue and each summer I made it a point to leave Washington for a few weeks and put one of my staff in charge. These short periods of time in the field were very profitable for me intellectually and some of my most productive years were while I held a position in administration. One of the reasons this was possible was that our Director, Dr. Thomas Nolan, was a research geologist himself and he also took some time in the summer to carry on his geologic mapping.

The following sections describe a few of the projects that engaged my time.

The Concept of *Base Level*

During his famous expedition down the Grand Canyon in 1877, John Wesley Powell was impressed with the sequence of rocks in Marble Canyon and Granite Gorge. He saw in the cliff walls nearly horizontally bedded sedimentary rocks of Carboniferous age lying on the nearly flat surface of erosion cut on folded metamorphosed rocks of far greater age.

He could see that the old underlying rock had once been not only folded and metamorphosed, but had once been raised up to the surface, where over a long period, it had been weathered, eroded, and reduced to a nearly flat plain. This erosion period had so reduced the surface that there was no longer any appreciable slope for further erosion. The erosion had reached a base that he described as *base level.*

> We may consider the level of the sea to be a grand base level, below which the dry lands cannot be eroded; but we may also have, for local and temporary purposes, other base levels of erosion, which are the levels of the beds of the principal streams which carry away the product of erosion.
>
> What I have called base level, he continues, would, in fact, be an imaginary surface, inclined slightly in all its parts toward the lower end of the principal stream draining the area.

This simple and useful concept was immediately accepted by a large number of geographers who also expanded its application in several ways. This growing complication was continued 25 years after Powell by the physical geographer who dominated the field for the first quarter of the 20th century, William Morris Davis, professor at Harvard. He summarized these various usages and their variations in a useful and forward-looking essay, but it is interesting that none of these many authors dealt in any detail with the cases of a rise or fall of a temporary base level. Though these may seem like minor complications, they have some very practical and perhaps unanticipated results. From field experience I became concerned with these matters.

I have described earlier the experience on the Navajo Reservation and the discussion with an engineer concerning the deposit behind a check dam. I had seen that deposition did not lead to progressive filling upstream. As years passed and my experience broadened, it was clear to me that the dam in a stream was a temporary base level as described by all the previous geographers, but none to my knowledge looked at the details of the various cases of a falling base level, a rising base level, or the results of changed watershed conditions when the base level was constant.

Early in our joint work on stream characteristics in the semiarid west, John Miller and I decided to observe the development of deposition behind a dam, so we built in each of two small arroyos a dam made of concrete blocks with mortar between. The lip of the dam was a couple of feet above the original streambed. Upstream we put in closely spaced pairs of iron benchmarks between which we surveyed a series of cross-sections. Our intention was to measure the deposition at the end of each season. These surveys were repeated for several successive years.

From previous observations we expected the deposition behind the dam would take several years and indeed this was observed. The head, or upstream position, of the wedge of deposition moved upstream until the surface slope of the wedge became about 50% of the slope of the original channel and further deposition ceased. All succeeding sediment passed over the dam. This last statement posed a problem.

The movement of sediment by flowing water has now been studied for nearly a century. It is now agreed that this movement involves the expenditure of mechanical power just as it is required to propel an automobile or any other machine. In hydraulics it is the product of the discharge and the slope multiplied by a constant, the weight of water per unit volume. In the case of the flow over the deposited bed the slope of which is only half the slope of the original channel bed, how is the sediment transported at the original discharge but half the slope? Before trying to solve that hydraulic problem I wanted to study field examples in which base level had been changed or unchanged but the inflow continued to be the natural runoff of the basin. And I wanted to prove that the sediment deposit did not increase with time after it had established its new equilibrium at the reduced slope.

In our flying experience we had seen several instances when, in canyon country, a meander had cut off and isolated the bend from the river that continued its progressive downcutting. The great canyons of the Colorado system developed as a result of gradual uplift of the land by tectonic forces and the rivers eroded downward to match the gradual rise of the land. This process led to an isolation or vertical difference between the river and its former level.

The Loop of the San Juan

Of these isolated river bends, the one in the San Juan River in Utah called *The Loop* provided a perfect example of a cliff separating the present river from its former bed. The Loop had been a meander in the channel during the early part of the Pleistocene, probably 1.5 to 2 million years ago. Charles B. Hunt states that the bottom of the Grand Canyon was within 50 feet of its present depth 1.2 million years ago. The base of the gravel on the bed of the ancestral river is 170 feet above the present river so the meander cutoff must have taken place over a million years before present. The earth's crust continued to slowly rise and the river cut down as a result, but the abandoned meander was isolated from the river with the result that the open end of the cutoff valley became a waterfall down to the river.

Rain, always episodic, fell into the isolated valley and drained off over the developing waterfall. The cliffs of the isolated valley some 1,000 feet high slowly eroded, the debris from which collected in the bottom, and some was carried to the lip and over the fall. But to move that debris, sand, silt and rocks toward the overfall required a slope. The accumulated sediment built up, the deepest part being at the apex of the curved valley, so two separate drainage routes developed, each sloping from the apex of the curve to the overfall. The accumulated sediment finally achieved a slope toward the lip sufficient to carry debris from the slowly eroding walls, and an equilibrium was established between rate of sediment supply and rate of transport toward the waterfall. How long this took is unknown.

We surveyed the entire length of the curved valley. At the apex of the curved channel, the deposit was 190 feet above the bed of the ancestral river. On the second expedition to work on the Loop we had a canoe carrying my wife, Barbara, and me and a very small rubber raft with Bill Bull and Will Haible. We started near Bluff, Utah, at a very low flow of the San Juan so the float down was difficult. The rubber boat continually shoaled and was leaking; it was a disaster, but the weather was perfect and the scenery superb. The spectacular and seldom-seen petroglyphs on the red sandstone cliffs could be reached only by boat. The rocky shores near them were ablaze with blooming cactus.

We made a simple camp at the base of the overfall cliff where a narrow sand beach had enough driftwood for fires. It was a dream of a camp. I sat in the stern of the beached canoe and caught catfish that I skinned and cut two into thin fillets that Barbara cooked over a sizzling grill. These were some of the best meals in all my camping experience. To get to work we climbed the cliff above camp and began our extensive surveys.

The water flowing over the bedrock lip of the overfall had gradually eroded the lower reach so the present lip is 70 feet lower than the bed of the ancestral river. A great deal of geologic work can be done in a million years.

The lip of the overfall was the temporary base level for the ephemeral streams draining the valley. These streams were stable during periods of relatively constant climate but there were periods when, owing to climatic change, sediment production from the eroding cliffs increased and large amounts of sediment accumulated in the channels in the process of aggradation; but at the outlet the elevation of the new deposit was not controlled by the cliff. Within a few hundred feet of the lip, the extra sediment just dumped over the cliff as if it didn't know the cliff was there, so, as can be seen in the profile, as much as 20 to 40 feet of sediment was deposited and the thickness was not controlled by the lip or the base level.

Later, climate changed again and the new deposit was trenched by a gully and left as a terrace. The same thing happened a second time, a new lower level was built, then gullied, and a low terrace was left as remnant.

The geomorphic conclusion is that a nearly constant or very slowly falling base level exercised its control of deposition only over a very short distance. This conclusion is reminiscent of the earlier thought that lifting the local base level by a dam would affect the channel upstream for a long distance. Observation showed that to be erroneous, for the effect upstream extends only a very short distance.

Mesa Verde

Another example of the effects of climatic change on a basin with a cliff providing a constant base level I saw in the geomorphology of the surface of Mesa Verde, Colorado, a relatively recent archaeological site famous for its ruins. The Paleoindians lived on the top of an isolated tableland, the surface of which stood some thousand feet above the Mimbres River and its drainage. The nearly flat surface of the mesa is forested with many grassy openings cultivated by the native population. On this tableland climatic change altered the surface channel system as it had elsewhere in the southwest. Deposition occurred and later gullied, a sequence common in the region, and, as in the case of the Loop, two terraces developed. The outlet of the channels ended at the lip of a vertical cliff down to the river. The sequence of deposition and erosion occurred in apparent independence of the fact that the channels ended at a cliff. The deposition of the high terrace was tens of feet thick and abruptly ended about 100 feet from the lip of the cliff. The local base level exerted its effect a very short distance.

But the Mesa Verde case provided even more information on base-level effects. The gullies that were cut in the deposited sediment deprived the inhabitants of some the land they had cultivated, so they built check dams in the gullies, behind which the deposit formed a small flat but stable area to plant crops. The pollen analyses we made showed they were growing corn as well as other plants. The people learned from experience what some modern engineers did not know: that a raised local base level caused deposition only a short distance upstream.

The Negev, Israel

There still remained the problem of what happens to a raised-base level over long periods. I went to Israel to study the dams built in dry valleys of the Negev by the Nabateans, who lived in a large area of what is now Israel roughly between 200 BC and AD 200. These dams were made of rock without mortar, stood about 4 to 5 feet high, and extended across the wide valley over 200 feet or more. In this low-rainfall area it probably took several years for the sediment wedge to be extensive enough behind the dams to be ready for planting crops. But these people learned that the alluvium would provide good soil material for planting and would utilize the runoff water effectively as it flowed over the low gradient deposits from one dam to another. They knew how far upstream the sediment wedge would extend and they spaced the dams so that the upstream edge of the wedge would be at the base of the next dam. In one little valley there were 13 successive dams.

At the time of my survey, a local Bedouin had planted wheat on the deposit behind the dams I was investigating, so that these structures are still in use after two millennia. He sent a message that we were hurting his crop and when I heard that I sent the one student who spoke Arabic to say that we were sorry and we were immediately leaving. In reply he asked me to have tea in his black hide tent. I turned it down with much regret because I could not ask my students to wait while I had a social visit.

Effect on Geologic Thought

The results of all these investigations affect some basic ideas in geomorphology. Two centuries ago Playfair observed that, when a tributary meets a master stream, its bed elevation coincides with that of the master. In this he was correct. However, people later extended that basic truth in an unfortunate way. From Playfair they made the assumption that whenever the master stream degraded or lowered, it forced every tributary to do the same. We showed that lowering base level had such an effect only a short distance. A more appropriate interpretation in my opinion is that whatever climatic forces caused the main stream to lower, affected tributaries in the same way and caused them to lower. It was not a base-level effect. But more important and more general is the claim by Prof. Hoover Mackin that the adjustment to a new condition takes place by a change of bed slope to provide "just the velocity required for the transportation of the load supplied from the drainage basin." He went on to state that his study of the subject "tends to confirm the standard geologic view that streams readjust themselves to new conditions primarily by adjustment in slope, and only in minor degree by modification of channel section." In fact, our work shows that slope is the most conservative and nearly invariant of the hydraulic parameters. The adjustment caused by a change in relation of discharge to load is made through the alteration in pattern, in roughness, in morphology, and in width and depth. This difference materially affects our understanding of grade, that is, the quasi-equilibrium in channels.

In stating his conclusion, Mackin was influenced by the famous experiments of G. K. Gilbert, but did not understand the implications of difference between the laboratory technique Gilbert used and the field situations to which the concept was being applied. Gilbert's flume had a flat bed without slope and as the sediment in the water was deposited it gradually built a slope toward the end of the flume until further sediment traveled the length of the flume and fell over the end, exactly the situation in the Loop after it was abandoned. Sediment deposited on the flat floor developed a slope sufficient to carry all the sediment with no further slope change. After that any excess or deficiency in sediment aggraded or degraded the channel bed with no change in slope. The Loop is an infrequent event in geology. The usual case to which Mackin referred is illustrated by the Mesa Verde situation and the Loop after it had developed its equilibrium slope. Both aggraded and subsequently degraded in repose to change in load without change in slope.

Here was a case that a hypothesis drawn from an empirical observation was never subjected to a test that may disprove the hypothesis. When such a test was made, the hypothesis failed.

River Meanders

When one looks out of the windows of a commercial airliner flying at high altitude, the one feature that is recognizable is the silver ribbon of a river far below. The river may be straight or in many reaches curls in attractive loops or meanders—at times it is amazingly repetitive. Though the casual observer probably would not notice, the shape of these meander curves is similar regardless of the size of the river. In fact, if one did not know the relative size of the river seen, he could not tell from the shape of the meanders whether a large river is being seen from afar or a small river is being seen close by.

Because the meander curves are so common their origin is a major problem in geomorphology and geography. Many investigators have written papers on the subject over a period of years. We found that it was not possible to attack this problem directly for there was no obvious reason, no logical hypothesis to guide an inquiry. In such a case it was logical to study one minor part of the problem at a time, observing that element in many rivers and quantitatively recording the results. This was an approach and it

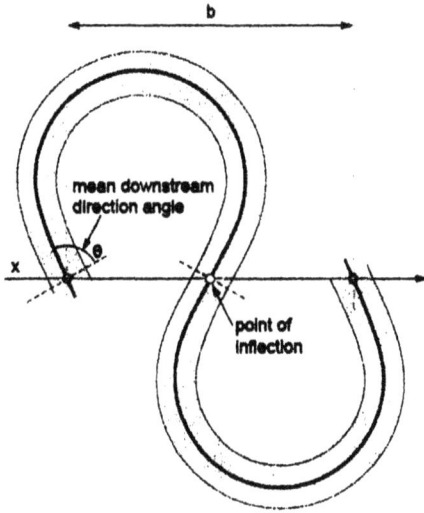

Meander geometry.

consumed part of my time for many years. It was a major undertaking for the knowledge and understanding increased only bit by bit.

The first stage was devoted to describing the curves in quantitative terms. It did not take long to see that the width of the channel was closely associated with the pattern of the repetitive curves, and after studying many rivers I was able to show the width was related to the wavelength of the repeating pattern. The wavelength is just about 11 times the channel width in natural channels. This led naturally to the measurement of the radius of curvature of the loops and we found that the wavelength, indicated by b in the adjoining sketch, is 4.7 times the radius.

We also found that there was a consistent pattern in the deposition of the coarser sediment at the inside of every bend in a form called the *point bar*. This deposition on alternate sides of the channel occurred whether there were well-developed curves or where the channel was merely sinuous. In fact, the alternate deposition occurred along perfectly straight reaches of river when they occur, for straight segments are amazingly rare.

It was obvious that when a river meanders in a valley, it is taking a longer course than a straight channel to flow the same distance down valley. Hence the mean slope of the meander is less steep than for a straight reach. If the mean velocity is to be maintained while the slope decreases, because the resistance factor is inversely related to the velocity, the roughness must be increased. This increased resistance is apparently provided by the meander curves. To state the same in another way, the introduction of curves introduces more resistance to flow than would be present in a straight channel.

The effect of curves on resistance to flow had been assumed but never specifically measured, so we designed a set of experiments to measure the resistance. We constructed a flume in the laboratory of the University of Maryland, hinged at one end so that when the other end was lifted or lowered the slope would be changed. The flume was filled with sand about 6 inches deep into which we would scrape a channel of constant width and depth but curved into meander form of chosen dimensions. Different amounts of curvature were carved in the sand and the measurement repeated.

The experiments showed that the resistance to flow increased as the amplitude of the bend increased and that only a modest increase in amplitude caused resistance far in excess of that caused by skin friction on the bed and sides. This was an unexpected result. We also found that a new and unknown type of resistance to flow that we named *spill resistance*.

But these approaches are insufficient without extensive fieldwork. I chose to work in detail on one of the most perfectly formed meanders I ever saw on the rivers where I have worked. They are located on a beautiful flood plain with cottonwood trees and large areas of grass on the Popo Agie River near Hudson, Wyoming.

For several successive summers we camped on the lovely flood plain with the canoe beached near our campfire. Along this reach we established a series of cross-sections monumented at each end so that we could repeat the measurements. A major purpose after surveying the form and profile of the bed profile of the reach was to establish the profile of the water surface. Each summer the discharge was different so over successive years we had a set of profiles of different water heights. These showed in detail how the profile is nearly flat around each curve, but steepens slightly at the crossover, or the place the curvature changes. Further we found that the higher the water surface the less difference in these parts of the curve, and near bankfull stage the water surface becomes a straight line.

On our last trip to this favorite spot we experienced a long and heavy rain; the water rose to the level of the flood plain, then inundated it and flowed through our camp at high velocity. In the dark we piled our gear on the air mattresses and floated them out toward the road finally swimming to get them on land above the flood.

On another trip we hiked into the Wind River Mountains toward the largest glacier in Wyoming, which lies just below the highest peak called *Gannett Peak.* I wanted to measure and photograph meanders of the meltwater streams that flow across the top of the glacier ice and gradually melt their way into the ice as deep meandering channels.

During another summer I had the opportunity to go to the Aletsch Glacier, located near Brig in Switzerland. This is a monster piece of ice that is miles long and thousands of feet deep. It made little Dinwoody Glacier in Wyoming a toy. With the help of my Swiss guide I made some useful observations on the meandering meltwater streams on the glacial surface.

The importance of these observations on ice is that meanders occur in the complete absence of sediment load. Later I found in the literature that the Gulf Stream off the coast of southeastern United States meanders, and these occur in complete absence of banks that bound both rivers and ice streams.

So we were at a point where a great deal was known about meanders, their geometry, occurrence, relations to bankfull stage, and effect on flow resistance, but we still had no general theory. I felt there was still more to learn from channels that do not meander and I wanted to find a way to investigate that.

I had seen a small stream in Wyoming that had some beautiful meander bends not far from a rather straight reach. I decided to study this place. It was early summer and the snow in the mountains was melting fast so the valley streams were running high for a few hours each day of melt. This stream was far enough from the mountaintops to be bankfull about midnight. Not deterred, my associate Bob Myrick and I flew out to Wyoming, traveled to Baldwin Creek, and set cross-section stakes all around one meander curve.

About 100 yards downstream was a relatively straight reach for which we also set cross-sections. Here we were ready to compare a full curve and a straight reach, with the same flow, the same sediment load, the same bank materials. Our main problem was that bankfull stage would be at night, after our extensive mapping during the day.

We lit a lantern, got out our chest waders, and began surveying. The water was chest deep, the width some 20 feet, the water up to the top of the banks. At each cross-section we measured the water surface elevation, the vertical profile of the velocity as well as the cross-section of the bed. It took all night with some frightening incidents when the velocity nearly pushed the rodman off his feet. The level rod was illuminated by the lantern at the instrument man, but the level instrument was put close to the bank so the distance to the rodman was not too great. Though the rod was sometimes hard to read in such poor lighting, we got it done. The quantitative data for that investigation occupy 110 pages in my field book.

As daybreak came, we got out of the river and to our simple camp where we dived into our sleeping bags. A few hours later I built a fire, got some coffee going, and sat down with the notebook to plot the data. About an hour went by and I kept looking at the graph I had just plotted.

"By God," I spluttered. "The meander is steeper than the straight reach!" I looked at it in more detail.

"The profile of the water surface in the curve is straight, but it is steeper than the non-uniform profile of the straight reach. What the hell is going on?"

Then I said to myself that the river is trying to do something. It is somehow compensating for something. The profile of the water surface in the straight reach is no longer stepped but practically straight, so it looks as if the river is making a straight water surface profile though it is slightly steeper.

Later we realized that this is exactly what was occurring, for the river gained in uniformity (straight profile) at the expense of some energy loss (steeper profile).

This was the most exciting moment in my career for I had stumbled on a completely unexpected reality in the field and the role of the scientist is to find the theoretical reason for the observed fact.

It became clear that the steepening of the curve came about by increasing the slope of the pool reach that at low flow was nearly flat. As a result, the total or mean slope of the whole curve was steeper as discharge increases. The break in slope from the pool to riffle was gradually eliminated as the profile of the water surface became uniform with increasing discharge.

Because the water surface slope is a direct measure of the rate of energy expenditure, the increase in discharge to bankfull stage leads to a greater uniformity of rate of energy expenditure but this is accompanied by an increase in the total energy loss. The idea of a mutual compensation was an entirely new thought to me and it resulted from a simple experiment that yielded improved understanding.

Because the adjustment described is local, that is, within a channel curve, it does not affect the obvious relation that lengthening the path between two points decreases the mean slope between them. In the case of Baldwin Creek, my survey extended through a distance of more than 4000 feet in which the slope down channel was 0.0056 as compared with the downvalley slope of the flood plain, namely, 0.0120.

On my return to Washington, I discussed the findings with my close colleague, Walter Langbein. We decided that the theoretical reasoning we needed lay in the concept of entropy and we soon discovered that the theory elaborates just the kind of occurrence observed; there are opposing tendencies for uniformity of energy expenditure and minimization of energy lost, well illustrated by the field case just described. With this concept in view many other correlative features came to view, especially the concept of the minimization of deviations or minimum variance as in statistics.

Now we realized that we should be looking for the minimization of deviation from some mean. In this view, a curved channel was continually varying from a straight downvalley path, therefore the deviations of direction at any locality on the curve from the downvalley direction was an example. So we measured these deviations at successive places along a meander and computed the sum of the deviations. Then the square of this value represented the variance as in statistics.

I repeated the process on different curves of the same wavelength and amplitude, a half circle and a parabola. The result was that the sum of deviations was smaller on a meander curve than on the other types of curve. This meant it was more probable. We had as yet no quantitative description of the meander.

In the literature we found that a contemporary scientist Dr. H. Von Schelling had studied the curves of a column of smoke rising from a candle into still air. He wrote an equation for the curves. Walter was able to simplify his equation to a form relating the deviation from a straight line to the sine of the distance along the curve. Thus, the angle of deviation from a straight line plotted against distance along the curve was a simple sine curve. We then named the meander curve a *sine-generated curve*, a distinctive type characterized by minimum deviation.

It was clear from the theory of entropy that there are two tendencies, the one for minimum expenditure of energy, and its corollary, the tendency for uniform rate of energy expenditure. These cannot both be satisfied, so the result is a compromise between the two. The longitudinal profile of a river is an example of this compromise. Uniform rate of energy loss would result in a straight profile of a river from headwaters downstream. Minimum total energy utilization would mean a very steep headwater rapidly flattening to a very gently slope near the mouth. The profile of real rivers is about halfway between these extremes.

So we presented a hypothesis, the theory of minimum variance, that explains a variety of the observed characteristics of rivers. No alternative hypothesis has been proposed nor has evidence been presented that refutes this hypothesis.

The search for an explanation of river meanders is an example of how the inductive method is used in geologic–geomorphic studies. Observations in the field and laboratory illuminate small aspects of the main problem but prepare the way for a breakthrough when finally some element gives an insight into the larger issue.

The Grand Canyon

The day came in 1965 when we were equipped with two rubber rafts, 22 feet in length, each equipped with an outboard motor. The spring trip on the Colorado would be on a discharge of 50,000 cubic feet per second, greater by 15,000 than the maximum discharge possible from Glen Canyon dam. As it turned out, this was the last trip down the river before the gates of the dam were closed, so we were to be on the river under conditions quite like those experienced by Powell on his original expedition. My main purpose was to see if it were possible to make a complete record of the water depths through all the canyon sections. It would also be the last time this was going to be possible and probably the last time the discharge would be high enough to represent the usual spring condition of flow. When the gates are shut, the amount of water that can physically be allowed to pass through the outlets will be far less than that of the week we had chosen to start.

There had been one previous scientific expedition by the US Geological Survey in 1923 run by the early river explorer Emery C. Kolb. He had, of course, used wooden boats, as did Powell. But for our trip we needed a platform from which to let a 50-pound lead weight down with the current water to measure velocity. For this purpose, one boat was fitted with a wooden deck the full width of the boat and 9 feet long on which we fastened a winch to pay out the cable carrying the meter and weight. We carried an alidade, a plane table, and a tripod, two depth sounders that were non-recording but that could read depth directly. There was a radio to reach an airplane flying overhead. We needed trailers to carry the boats, trucks to pull them, and drivers for the vehicles. These were procured by the research office of Herb Skibitzke. Some years later I floated in a wooden dory down both Cataract Canyon and the first half of Grand Canyon but a boat like a dory would not have been satisfactory for the kind of observation and measurement we wanted in this expedition. To prepare for this expedition Howard Chapman flew the whole route taking photographs from a very moderate altitude. Chappie was a former fighter pilot in the Pacific theatre as well as an experienced field man.

We assembled on the San Juan River at Mexican Hat, Utah, in early June. There were two boats. The personnel consisted of the boatmen, Smuss Allen and Bruce Lium, who was also a limnologist; Chappie, the engineer; my sister, Estella Leopold the palynologist; Meyer Rubin the geologist and expert in C14 dating; Emmett and Myrick; and some helpers. After Cataract Canyon, Herb Skibitzke said no more boats for him. He'll do the flying.

We started down the San Juan, which I found surprisingly swift. The surface velocity was 5–6 feet per second in the pools and nearly twice that in the rapids. Depths averaged 4 feet but increased as we went farther downstream, presumably due to gradual narrowing. The aerial photos were a tremendous success for they showed clearly every big rock and one could see exactly where the boat was at any time. The photos were printed on very long rolls, and each photo overlapped the previous one; we unrolled the photos as the boats moved. One person constantly watched the visual signal of the depth on the depth sounder and called out the depth at least every $1/10$ mile and all depth changes in between. Another man constantly looked at the photograph to keep track of our position and every time the recorder man called out a depth, the photo man wrote that depth down on the photo at the spot the depth was read.

We explored one tributary valley that was filled with an old massive debris flow ranging from silt to 12-inch boulders mixed up without stratification. At Sulphur Springs there were two ages of talus. The older accumulated on bedrock so it was not destroyed by subsequent debris flow. Old talus tends to be dark in color and new talus a very bright tan. We measured depth, cross-section, velocity, slope, grain size, and profile. Velocity was measured by floats.

The camp made 200 yards below Sulphur Springs was as beautiful as I could imagine. A spacious rocky ledge occurred a few feet above the water surface and was free of sand. Facing camp was a sheer red cliff to provide scenery during the day, a silhouette at night, an edge over which the moon climbed, and shade for breakfast. The pool of still water below a red polished shoreline provided a fishing hole for catfish. That night there were drinks and a guitar in front of the fire.

And so one interesting day after another. At a place called *Indian Meadows* the San Juan becomes so shallow on a sand bed that our boats were stranded and all of us pushed until we finally floated again. In the midst of this the plane came by and dropped water and ice, but the water had been put in a washed gasoline can and so was unusable. Learn something every day. Smuss told us that this shallow sandy reach prevented the San Juan from being used as a river trip for tourists.

We came upon the beginning of the delta of the San Juan emptying into Lake Powell. The water became merely 2 feet deep for some miles and finally pitched off at the avalanche front of the growing delta. There the water suddenly becomes clear instead of reddish dark with silt. It was necessary to cruise down Lake Powell about three quarters of the distance from Rainbow Bridge to the dam. There are beautiful little coves partly submerged and full of bluegills and bass. At several places small springs drip water over red cliffs and moss, ferns, oaks, and poison ivy are plentiful. We visited Rainbow Bridge, a rather sad experience when one thinks of its former wilderness state—so distant that few people ever saw it. The environment changed radically when we came to the place where we could take the boats out of the water for the portage around Glen Canyon Dam. At this place there were lots of people, a pair of unclean latrines the Park Service had installed, full trash cans, and dirty fireplaces. When we saw this we did not land but camped some distance away at the mouth of Bridge Creek on a low sandstone saddle. The next day we were met by our helpers who had come with trailers on which we loaded the boats. It is a rather long distance by road to go around the dam to the vicinity of Lees Ferry, where we launched.

The first day on the river in Marble Canyon was different from anything I had previously experienced. The discharge was 50,000 cubic feet per second. When I measured the width by plane table at Jackass Canyon, it was 400–450 feet, the velocity average 5–7 feet per second, and in the rapids 11 to 13 feet per second. The depth averaged 25 to 30 feet. At 5:50 in the morning the sun was making gold on the upper cliffs, while camp and the walls about it were still a special red chocolate and purple. The rapids seem to both talk and roar at the same time.

Some miles of Marble Canyon are sheer with but little talus. In other places the talus slope reaches up to 200 or 300 feet on the canyon wall. In the larger rapids the fall in elevation takes place mostly in the upper part of the rapids before the smooth slick turns to waves. At one rapid I measured the fall of the water surface. It had a vertical fall of 13 feet in a distance of 300 feet.

A party of two boats tried to join us—apparently even camp with us. Our crew felt I was pretty cool to that idea and indeed I was. I felt it was the same as some stranger asking about your favorite fishing hole.

The water temperature was 56 degrees but it seemed cold when one bathed. We camped on the left bank in a small sandy recess where there was plenty of driftwood for a fire. I found a scorpion that was 4 inches long with a purple body and yellowish legs, head, and tail.

At Triple Alcove I was sitting against a rock on the boulder fan looking upstream at the west wall flooded with some light while I was in the shade of a cliff a thousand feet high, perfectly vertical except for two small talus-covered ledges at the base. The wall was an exquisite rose tinted in browns, grays, and greens. In the sun the wall was a buff rose. The river reflected the green of the lower slopes and yellow with a touch of the rose.

The day before we had walked up a dark narrow side canyon on terraced steps of white limestone where there are seeps supporting maidenhair fern, thistles of great size, redbud that usually occurs only in the eastern and far western states. There were also some very tall cottonwood trees. At one place the narrow notch gave way to a great theatre at the far end of which a rock terrace protruded as if it were the stage for this theatre. Around the border walls there were grotesque erosion forms as if there were thousands of people crowding its balcony. On the stage the men could not resist declaring in oratorical form.

On another excursion Bill Emmett and I climbed above a sheer wall where there was a little spring. From this height we could look down at a perfect meander cut in the rock, only 3 feet wide at the bottom, and, though flaring upward, the meander form had been preserved for the last 50 feet by downcutting in the rock.

Just above Vasey's Paradise I was fascinated by the polish the silt-laden water had produced in the walls up to about 6 feet above the water present water level. This wear by the water left lovely facets, hollows, and grooves in the limestone. I kept wondering what length of time was required to do this, and similarly the time it takes windblown sand to facet the ventifacts we have seen in Wyoming and elsewhere.

The airplane dropped a parachute carrying some rope I needed, some guitar strings, and water. The drop landed high on the talus half a mile upstream. Just above camp was the wreck of a wooden boat that was the end of a man who didn't make the big rapid we hit the previous day.

There were gravel terraces standing 200 feet above camp. We made a long stop at the mouth of the Little Colorado River and we measured the discharge some distance above the mouth. We understood that it was only the 6th measurement ever made at the mouth of this important tributary. The main point of measurement is at Cameron, many miles upstream. At the mouth the water is bright turquoise in color from, I suppose, copper sulphate. On the main river I made a measurement by floats.

The depth-measuring device on one boat was working perfectly but we broke the transducer column off the other boat when it hit a rock. The combination we had devised of unrolling the aerial photos and writing the depth on the photo was working perfectly. Considering that this was the last river trip anyone would make before Glen Canyon dam closed its gates, the data we were obtaining at 50,000 second feet would be unique. We were getting a depth measurement at least every one-tenth mile plus the many at intermediate points.

The air drops by parachute with which we were experimenting was well worth the trouble despite the fact that the local winds in a deep canyon can blow the parachute off course on very difficult terrain. Before the trip I had asked the crew not to drink water from the river and the fresh supplies were quite needed. Only at Deer Falls and at Bright Angel did we get fresh water locally.

A most impressive observation was on the gravel terrace that we kept seeing, standing 200 feet above present river level. The weathering on those rocks in this truly desert climate was unexpected. Nearly all the big limestone rocks are rounded, and deeply reduced in size. Solution facets of great variety were seen. Where a hard inclusion, such as quartz, occurred in the limestone, it stood out as much as 4 inches. This means that the rock had that much dissolved away during the time it took the river to cut 200 feet in this arid climate. It is arid here and hot, supporting cactus in profusion and mesquite. The trap rock is shiny black with desert varnish.

At Elves Chasm, we were out of the limestone of Marble Canyon and into the intrusives. Four thousand feet of nearly sheer cliff faced the sun and our camp on the left bank. There is a small granite bench 30 feet above the water, graced by a little rivulet of water that drops over water-smoothed rock into a big pool. Above the nearby high cliff there is still a higher cliff, and another thousand feet into the sky there is the lip of the plateau.

The De Haviland Beaver flew over early, and in anticipation, several of us climbed onto inner hills where the drop was supposed to fall, but one parachute failed to open, some of the water cans wrecked on the rocks and another carrying, among other things, lettuce, provided us with self-shredded salad. One drop had a reel and a cable for the current meter that we retrieved intact.

At Unkar Rapid I decided to take some detailed measurements and set up the plane table and alidade on the shore to observe the stadia rod erected vertically in the boat. At different distances from shore, Smuss held a boat facing upstream against the current with the motor screaming at full throttle to keep it in place while the current meter was lowered, but it was all he could do to keep the boat holding its own. The current meter readings at different depths and at the surface were obtained using our heaviest lead weight on the cable.

The depth changes radically along that reach. At the lip of the rapid, it reaches its minimum value. Most interesting was the fact that the velocity measured by current meter at the surface and at the bed was the same, 11.5 feet per second. This tells one something about what happens in a big rapid. The water goes over the lip where it is most shallow, then plunges downward into the hole below, creating a strong bed current. Owing to the shear of the main current against the bed, there is formed an upriver current that concentrates near the banks and converges with the main current, thus forming horizontal whirlpools as well as strong upward-directed cells and random vortices.

To check some of this conjecture, I put a dye float just past the lip of a rapid where I felt it would dive downward. I stood up in the boat looking for it to appear again at the surface somewhere. This search continued for several minutes, much longer than I expected. To my surprise I saw the float surface about a quarter of a mile downstream.

A day later we were at the head of Granite Gorge and, after running Hance rapid, made camp on a large bar just above Bright Angel creek. This was close to the trail from the south rim and near the residence of the USGS hydrographer Bill Burnett. I inspected the cable across the river from which the measurements are made and crossed on that cable to the left bank. There I saw a very old wooden box that was long ago used to carry a current meter and weight.

That evening Herb flew in using one of our helicopters, landing it on the same bar as our camp. It was impressive to see him confidently coming down into that narrow canyon that seemed hardly wider than the span of his rotors. He still could not join us because there was illness in his family. We discarded all the water jugs that had been tainted with gasoline and filled the newer ones Herb had brought. His visit was also welcomed because he replenished the supply of Jim Bean. The advantage of having support by air was being amply demonstrated.

From Bright Angel we traveled through the narrowing gorge, in places a mere 150 feet wide, and through nearly continuous rapids, separated only by boils and whirlpools. Some of these boils surfaced a foot above the ambient flow, proof of a very strong upward velocity, frightening and awe-inspiring. The hard rock, gneiss, granite, and schist, had been polished by the silt-laden water over the centuries and presented a smoothly fluted surface, shining black in the sun from desert varnish. The fluting must mean times when high water worked on the sidewalls, whereas desert varnish represents periods of emergence from the flow.

In the 30-odd miles between Bright Angel and Elves Chasm there are some of the biggest rapids we encountered, especially Horn Creek, Granite, Hermit, and Waltenberg. Smuss knew the names of most of them.

At Elves Chasm we first saw the horizontal contact between the folded Precambrian metamorphics that were covered by flat lying sediments of younger age. This contact gave Powell the idea of peneplanation, the erosion of mountains until they are completely gone and only a flat surface remains, later covered with sediment layers on top of the old surface.

Lava had poured out of vents and formed a dam across the canyon, which created a great lake indicated by shoreline deposits high up on the canyon walls, as described subsequently by David Rogers among others. To us it was a matter of great interest to see the contact between lava and the native rock on average only 50 feet above the present water surface, the amount by which the modern river has downcut from the time of the great lava extrusion.

There was also evidence of an important period of deposition from springs in the shape of large masses of travertine, an evaporite from springs, hundreds of feet thick that hang like folded draperies on the canyon wall.

I took some climbing rope and with Estella went downstream to a narrow tributary where we could climb well above the channel and see that the water in this small canyon had cut vertically within a meander. The walls of this tributary were not polished by silt-laden water as is the main canyon, but had a patina that resembles a surfaced smoothed by fine sandpaper.

From Havasu Creek downstream to Lava Falls the canyon is particularly narrow with long relatively straight reaches and few sharp bends. The last big rapid is Lava Falls, difficult even at modest discharge but at 50,000, it is a real challenge. We got there in mid-morning, beached on the left bank, and all of us walked down to see how it really was. There are several big rocks, including one near the left bank at the head of the rapid that had to be avoided. The largest standing waves were just right of center. Lava doesn't look much worse than other rapids from shore level upstream, but when one climbs up on the left bank terrace and looks down at midrapids, the magnitude becomes apparent and more fearsome. When the others had turned away, Chappie and I were standing next to Smuss, who was unusually serious.

"Well, Smuss, how does it look?" I said.

"Well," he drawled, "we might make it."

"How do you want to do it?"

"Tell you, I'd like to have two volunteers go with me. If we make it all right, we'll be there to pick up the survivors. If we don't make it, there will be enough people to rope the second boat down."

"If that's the deal," I said, "obviously I'm going with you and I'd like Chappie with us."

"That would be my first choice," Smuss replied. "How about it, Chappie?"

"Of course, let's go."

The rest stayed onshore of the left bank. The three of us got the ropes ready and checked tiedowns and bucking ropes. I sat in the center, Chappie sternward with Smuss. We pushed off. It was as if we were in a dream charged with potency and dread. We braced and headed into the V of smooth water, just left of center. Going over the crest, the bow seemed to bend downward heading for hell. Throwing the tiller to the right Smuss gave it full throttle and the boat massively responded with a slight left-shore direction as we just slipped by the great hole behind the boulder on the left, just missing the edge of the deep vortex. There we hit the first big wave, which from the trough loomed ten feet over my head. The boat bent nearly double as the bow rose up toward the sky, then collapsed down as we went over the crest; the stern then lifted up to the wave top, the motor screamed as the propeller came out of the water at full power, while simultaneously the bow plunged head-on into the next even larger wave. Again the bow slammed down into the trough, and then screams from the airborne motor, and we hit the next wave. By this time the cross-channel motion was sufficient to have moved us just left of the next five big waves that towered over our heads just off the starboard gunwale. Then we could see that we had made it and we drew farther and farther into the great eddy of the left bank, motor still running full power as Smuss headed it upstream before we were caught in the next rapid downstream. The boat banged into a tall thicket of cane and I jumped, falling on my first step into a great hole of the uneven vegetation. Scrambling to my feet and giving a strong heave on the rope, I had it against the shore. Tears flowed down my face.

By the time Chappie and Smuss had clambered out, I had the rope secured around a big rock and tied. We walked together up the gravel slope. "No two men I'd rather have had for that one," was Smuss's only comment. In a sort of relief of tension, we all burst out laughing. I'm told Estella can't even talk about how it looked from the shore. Lava Falls at 50,000 cfs is one to remember.

We weren't through it yet. In the second boat, Bruce Lium took Meyer and Bruce. At least we knew it was possible. Smuss, Chappie, and I stood by in the first boat, motor running, in case they needed help. Bruce took the central V a little to the right of where we entered to give the big vortex a wider berth. This made him couple of seconds late with respect to gaining his cross-channel motion and as a result he hit the first big roller when it was even larger than at our position. So violent was the pitch when the motor came out of the water at the wave crest, Bruce was thrown forward on his head, leaving the tiller free. The motor torque threw the motor into a hard right turn and in the couple of seconds it took Bruce to recover and get back to the tiller the boat had taken a dangerous turn rightward and into the biggest train of waves we had skirted. The boat lurched violently into a sideway bank with the right gunwale very high, dangerously close to tipping over, but recovered with full power at hard left and within two more waves the boat was out of danger skirting the big ones.

"Every time it's different. If you make it right side up, it was a good run." That was Smuss's laconic statement.

After Lava it was just a river trip. There were impressive stretches of travertine hundreds of feet thick. Through many miles of canyon the walls gradually decreased in height. We went over a series of rapids, none of which compared with the many we had been through.

At the head of Lake Mead one is riding over the delta with only a modest depth but we became inattentive. Suddenly the bow went up radically and we began an astonishing roller coaster of very large waves that were high with steep flanks but not breaking at the top. This was a train of antidunes that are intermittent, move upstream, then subside, and in which the water crest coincides with the crest of the sand dunes below it.

This expedition made an important contribution to river science for it was the first set of measurements of channel bed over a long stretch of river length and certainly was the only one that included measurements through one of the great canyons of the world at a high discharge. Our measurements in the Colorado River and its main tributaries totaled about 6000 values of depth at recorded locations. It provides a quantitative evaluation of the frequency distribution of depth values from deepest to most shallow.

The Arctic Plain and Kotzebue

We had found air support a real boon, for photography, for reconnaissance, and for air drops of needed equipment. Herb and I started to talk about how geomorphologists could do new things if one flew and the other recorded and made notes and sketches. Herb had in his research group several pilots and he felt we should also learn to fly. And so it began. I was the first of our own scientific group who started to take lessons and even before I had my license I would help Herb by flying part of the time, but within weeks I was qualified and new possibilities were opened.

In consultation with Herb, Smuss, and Chappie, I laid out ideas for a river expedition down one of the rivers in the Brooks Range, Alaska. As such a trip required long range planning, Herb and I decided to fly to Alaska to learn about the country and the possibilities.

I flew commercial from Washington to Salt Lake City where Ray, one of our technical men, had pulled the De Haviland Beaver up to the gate. We left immediately for Great Falls where we were to meet Herb. I flew the whole distance and after a bit of waiting a car was obtained for Ray and Herb and I took off in the Beaver.

We left for Edmonton and Fort St. John from Great Falls a morning early in May. The last stop in the United States is at Cutbank, the next place we passed in Canada was Lethbridge. There are many prairie potholes in this region left after the glacier retreated, and the glaciated landscape has not advanced into an organized channel network. These potholes are extremely important for migrating wildfowl. The upland is imperfectly drained but has wide and open, trapezoidal-shaped valleys or gullies once eating headward.

The rivers are all cut down about 50 feet and bordered by high bluffs. The Athabasca is muddy like the Mississippi with large point bars of small gravel and sand. Then the country changed from a prairie of dark or black soil to a gradually more forested spruce–aspen association. The country came to resemble Russia near Moscow or near Sverdlobsk, a place of bog, white birch, aspen, and dark spruce. An amazing part of the earth is covered with this association, most of Canada and Alaska, northern Europe and Russia.

At Edmonton the customs officer looked at the cargo of the Beaver. He was much more friendly than the officer in the United States, who was surly at best. Here the officer explained that our shotgun was allowable but a handgun was not, so he sealed my six-shooter. The seals would be removed on our return trip.

At Fort Saint John we went about two streets from where we parked the plane and we were at the town center. The small hotel resembled buildings in early western towns with what appeared to be a false front but actually did have two stories. The weather was chilly and we wore our fur-collared parkas.

When we left Ft. Saint John we got out of the agricultural area and hit the spruce–aspen bogs. Rivers were wild, brown or copper colored or even yellow. This is especially true of tributaries to the Fort Nelson River. That river is a wild-looking torrent deeply incised in a gorge meandering between steep valley walls, many of which are perforated by slumps or landslides. There are numerous lakes.

It was interesting to see some details of at least one tributary to the Fort Nelson. There were lines of foam, beautifully distinct, that described the surface streamlines of the flow, emphasizing the crossings but never touching the banks of the meander. The same characteristic applied to the Mississippi: If a houseboat was tied to a floating log, the boat never touched the streambank.

The Peace River near Ft. St. John flows in a deep V-shaped gorge with bare valley walls. The Laird River is a gravel stream with large bars over which we flew up the channel only 200 feet above the water. It was most interesting to be this close to a big wilderness river and see the wildlife as well as the channel. All these rivers change character along their length. Most of this large area is in spruce–aspen bog but with no aspen. The spruce trees are very tall and sharp and parts of the flat bog have large polygonal patterns indicative of permafrost.

We left Ft. St. John and flew to Whitehorse, some 9 hours of flight. The sky was overcast, with high clouds and only a few moments of sunshine. Herb spent a lot of time keeping tabs on where we were on a map: The radio received nothing, the compass was useless, and everything looks the same. You have to know where you are. The biggest problem is icing that can build up fast as the weather changes.

The next stop was at Northway, Alaska, on the Tanana River. There is a combination airport–café–hotel–customs office–local store–bar. It consisted of a yellow log cabin mixed in with old, decrepit army barracks that were converted for the store. They had gas but no oil, so we dug out 3 quarts from out emergency supply in the Beaver. The drizzle alternated with snow from an overcast sky at 1300 feet. We sat in the hotel in a room that looks out over the decaying asphalt airstrip.

In this unprepossessing edifice there was an odor of overcooked eggs in bacon grease. People wandered in and out, helping themselves to a big urn of coffee. When we came in, the front hall was full of young Indians from the local village, population 300. These people apparently draw pensions as wards of the government and spend it here in the bar and in the trading-post store, located in the blackened barracks a few yards away. That store is quite like the trading posts on the Navajo reservation, full of canned goods; cheap clothes; bear traps; brooms; axe handles for big, probably dull, axes. The young Indians in blue jeans and cheap parkas were playing the pinball machine, listening to a bad imitation of the Beatles; one girl was imperceptibly weaving her hips to the music. The lady who is postmaster, bar keeper, hotel owner, cook, and customs official was telling about how the Indians are neither smart nor dependable. The day before she bought $150 of muskrat furs, price $1.50 each, that she will sell for a profit of 10 cents, but her store took in $1500 in a day. Her husband flew some Indians from SE to do biweekly shopping. So the feudal system of Hawaii, of the Tewa basin of New Mexico, and of the Navajo country, works here as well. You own the store so you can afford to help the locals and keep the money coming in. Muskrats are shot with a .22 rifle rather than being trapped. Bears are common, mostly Blacks but a few Grizzlys.

Leaving Northway, I looked out from the plane on this inhospitable landscape and contemplated the worsening weather. I kept wondering what we would find in the Brooks Range even before we reached the Arctic plain and Barrows. The character of the country is a glaciated terrain, mostly ground moraine that once was drained by meltwater streams long gone. These channels, devoid of water, are tortuously meandering on a very flat slope.

The weather report from the new weather station was hardly encouraging. Often the only forecast available is for the day, but what we needed was local and immediate. Between us and Fairbanks, the situation looked ominous. As we proceeded, it was sulky but passable, but near Big Delta it was snow–sleet with the temperature about freezing. That station is often in a drizzle even when everywhere else was acceptable if not cheery. I attribute this to a location in a narrow place between mountain ranges and at a high elevation. The cloud layer at 1500 feet looked thin so we picked a hole and climbed out of it, but at 10,000 feet on top there was only a white haze in every direction, and no horizon. The temperature was −5 to −10 degrees below freezing and ice began to form on the struts and leading edge. Herb worries about a lot of things but ice is a real first-class worry. He took the stick and made a steep descent through a hole when we could see the ground, and at 500 feet snaked our way along the river in a flurry of snow, drizzle, and fog. I tried not to contemplate landing in the unending bog below. It got soupy but the ice did not increase and 5 miles from Delta we called and they reported good ceiling and no rain so we went on, and sure enough, it opened up and we proceeded on to Fairbanks late in the afternoon.

The hotel we chose seemed to be the only establishment that looked something other than something on a frontier. The town seemed devoid of aesthetics or taste. The houses looked like shacks; there is evidently no concern with how things appear, there are no nicely kept yards or buildings, and junkyards of throw-aways seem to be everywhere.

The weather reports were not auspicious for Fairbanks or to the south, but Bettles showed clear though Ft. Barrow was in coastal fog. It did not look as if we would get a better day so we took off under a dull and bleak overcast as we headed for Rampart and Bettles, which lies just south of the Brooks Range. The sky cleared and the snow-covered peaks loomed above us. We got fuel for the Beaver at Bettles and headed north toward the Arctic. Barrows was fogged in and no one seemed to know what we might find at Umiat on the Arctic plain north of the Brooks Range. There is no radio aid anywhere north of Bettles so we were piloting on our own, the compass does not work at that latitude and one of us had to be constantly on the maps while the other flew the plane. Starting into the Brooks Range along the valley of the John River, the snow-covered peaks rose high above us, dazzling in the sunlight. This valley opens up into a large triangular amphitheatre and I had to choose which tributary valley to use. Only one of these ends up in a usable pass; all others end in impassable peaks. Following small-scale maps in this country is quite confusing because great valleys cannot easily be distinguished from small ones and I kept revising my estimates of where we were.

The John River is highly meandering. It is not very rapid, its water appeared brown from iron stain, and it has wide, gravel point bars. Those rivers without such bars would present impossible conditions for making a camp because the wet spruce bog comes right up to the streambank. This was particularly interesting to me for I had long hoped for a chance to take a wilderness river trip in the Brooks Range, and these gravel bars looked large enough to provide a landing place for the Beaver. I was beginning to formulate in my mind a test of the procedures we were developing for a quantitative description of a river. If we made measurements of the main channel and all its main tributaries during just a single trip down the river, how would this set of data compare with a long-term record at a few gaging stations?

Because we entered a narrowing V, I expected the actual pass at the top of the Brooks Range would be a low cliff or ridge beyond which we would go into a reverse sequence. Instead, the narrow valley gradually widened and was hemmed in by snow-clad peaks. Snow covered the whole valley and the mountains seemed to recede from the valley sides. We passed an Eskimo camp, Anaktuvik, a few tin-roofed shacks, much discolored snow, and in the vicinity dog teams pulling sleds. It was hard for me to believe this was the pass at the crest, but after a few miles I could see that the snow covered a creek draining north and winding in a flat trough within a broad plain of snow. This was on the north rim of the continent, an unbroken flat at high elevation but featureless. With this surprising development the tributaries we saw did not seem to form a coherent network and they joined in acute angles. Though neither radio nor compass was usable, we were clearly in the drainage of the Coleville River, which flows into the Arctic Ocean.

I was getting worried as the gas was being depleted and it was clear we must follow the correct branches of the river system or we would end up in a cul-de-sac. Meanwhile we saw many moose in the low brush near the channels, the snow was apparently not very deep, perhaps a foot or two. The sun shone on the white world in glistening brightness. Out of the corner of my eye, I could see a gray triangle over the sea of white, a V of three cranes, a perfect symbol of independence in this hostile environment, gliding in formation on outspread unmoving wings. These were the rare whooping cranes, white with the wingtips of solid black. On their navigation instruments they can wing it from the Arctic to the Gulf of Mexico and never get lost. We, on the other hand, had just heard our radio go dead and wished we had the same self-assurance.

I estimated time and ground speed and made notes of tributary entrances, trying to visualize how it would look on our return trip. If Barrow remained closed by fog, we must return the way we came. Though gasoline has been flown to Barrow for us, we were not to see it. In an hour or so we finally saw a big river, all snow and flat, just a large version of what we had followed. The Coleville River is bounded by low hills, or rather is incised in a wide shallow trough outlined by a few steep places when the snow was gone. In the triangle of three rivers joining, a derelict oil well stands stark, alone, and decrepit. Piles of oil drums and junk stick out of the snow. This was Umiat, an outpost in the Arctic, the shacks rusting in the sun; piles of discarded equipment stood black in the snow.

Moose tracks were to be seen everywhere. The gas gauge showed we must return. Barrow was still nearly 150 miles ahead. The sun had become obscured in a white-gray mist. Herb turned to me several times for assurance about directions. Once we found three tributaries coming together, all looking alike. He said, "Which one?" and I pointed confidently to the right side, yet as the plane banked I wished I could have the confidence of the cranes.

There were herds of caribou in the brush as well as in the muskeg openings on the benches above the river. When one drops down to see a moose, he wanders off unconcerned, but caribou break into a wild gallop in some direction.

We landed at Bettles with but little gas reserve. The weather looked ominous for tomorrow. We opened some C rations. I told Herb that if the weather got worse we would never see the Arctic Ocean. "How about going west to Kotzebue?" I said. He was game. After getting fuel we flew down the valley of the Kobuk River that flowed west like a small version of the great Yukon River.

Out over the intermediate delta of the Kobuk and over a frozen arm of the Arctic Ocean, we could see a narrow spot of coast facing west, piled deep with snow. The village of Kotzebue has about two streets, one being a track between piled snowdrifts, some 5 feet high. The snow was melting on the gravel tracks and, as in all towns, gets dirty. It was a messy mudpie over which we taxied the Beaver toward some likely looking gas drums. We were both tired from many hours of flying and I walked out for a stretch. A battered-looking and dirty station wagon pulled up near the runway and a grizzled man inquired what we were doing. He was from the Corps of Engineers, building a small dam for the Public Health Service to augment the local water supply. I asked if there was a trading post where I might buy some furs. Yes, they supposed I could get something that turned out to be the central general store. I was introduced to the manager who said they had nothing, but I kept pressing for information about who in the village was hunting seal. In the meantime the man who drove me went off to supper, it being 6 pm, but the sun was high in the sky. So I was left a long walk to the airstrip and I realized Herb did not know where I went. But I figured as long as I was this far, I would carry it through. Down the street (meaning a 6-foot strip of mud lined with snow drifts) was where someone else told me to look. I found the place but it did not look promising.

Richards, by name, second house past the red barn. "By the way," I inquired, "what tanned skins are these? Is this an arctic hare?" Pointing to a gray fur of small size. "Oh, that was shipped here from the southern states (I figured Arkansas) to sell to the Eskimos for their parkas." I had seen some real seal-skin parkas with a hood of wolf. But most are quilted cotton with a collar of rabbit or perhaps fox. This is analogous to the Navajo selling good blankets but wearing the cheapest cotton Pendleton of garish zigzag design. Before I left, Richards said, "Why don't you ask at Roberts down the street?"

I passed quickly along the narrow mud track, a few yards from the shore. The sea was dirty and mushy and on the ice was piled a great mound of offal only a few hundred feet offshore to be dropped into the sea when the ice melted. In the meantime in this warm weather it's a messy business. I jumped out of the way of a gas-driven caterpillar-tread cycle on skis, pulling a handmade Eskimo sled made for Husky power. This is the Arctic counterpart to the Totegoat of the western mountain trails.

I apparently found the right place and knocked on the door in the back that was answered by a large smiling lady who could be either Indian or Eskimo. In the room I could see about a dozen ladies sitting around a table just socializing. I asked the lady about seal skins. "Oh," she said, "my son just got back from harpooning seal and he got a few."

"The skins," I asked, "are some for sale?"

"Yes, we can spare one. They're outside."

We went into a small backyard and hanging from a clothesline were half a dozen hides, dripping water.

"Pick one out," she said. I fingered several and said, "But I don't want one that has those two long slashes in the hides. She smiled, "Those are where the legs were." I felt not merely like a greenhorn but a fool as well. She took one off the line and indicated we should return to the house. She spread the hide on the table as she recited to the ladies my inquiry and her answer. The peals of laughter could probably be heard to the North Pole. She fetched a carton of Morton's salt and spread it liberally on the flesh side of the wet hide and energetically rubbed it in, wrapped the hide in some paper, took my money and I departed with a red face but a treasure in my hand.

Somehow I found a ride back to the airstrip and rejoined a rather worried colleague, and because it was still broad daylight, we flew south. Our route along the coast was more weather-friendly than through the north end of the continent. At Edmonton the customs officer restored my revolver. At each stop I inspected the back of the Beaver where I had put the sealskin and to my continuing dismay, it leaked salty water all over the floor at the rear of the plane.

Finally we got to Salt Lake City where I called our associate Bruce Lium, and I urgently implored him, "Please get this damn sealskin to the tannery that you know and get it out of our plane and off our hands." The skin now lies on a bed in our Wyoming house with hides from Africa, Montana, and elsewhere.

The Mackenzie and Inuvik

There is so much to see that the next time we decided not to go to Barrow but to explore an area to the east, the Northwest Territory. We had installed in the Beaver a 50-gallon barrel for gasoline that through a valve could connect to the gas line of the airplane. We did not know where gas might be available because the chosen route was over even more wilderness than where we had been previously.

The first legs of the trip we knew, Great Falls, Cutbank, Edmonton, Lesser Slave Lake, then Peace River. From there north it was new and progressively more desolate with many more lakes than in British Colombia. We also wanted to make more stops to see the remote villages. Following the Peace River that flows into Great Slave Lake, the bog and birch forest seemed to alternate with forests of thin and dwarf-like spruces. We saw a female bear with a cub that shied from the airplane motor and she chased the young one ahead of her at a fast run.

At one place we landed on a very small strip, which was as usual close to the village. I saw some boys near the airstrip playing marbles. I wanted very much to know how these somewhat younger ones did, because the teenagers I had seen at Northway seemed to be up to no good, just hanging out at the store and bar. These youngsters were very willing to talk and I asked how many games they played in marbles. They showed me two and seemed glad to have me try one of their shooters. The first game was one with a large circle drawn on the ground with some marbles in the middle. A player apparently must shoot from outside the circle, with the result that the nearest one could be to the inner marbles was about 1 $^1/_2$ feet. There were other rules the details of which I now forget. The other was a game rather like one we played called *chase*. A player sent his shooter some distance and the other person tried to hit the first shooter, but if it was a miss, he would land some place rather far away. I found it difficult because I did not seem to send my shooter far enough so I was quite vulnerable. They got a kick out of being much more proficient than I.

We stopped at Fort Vermilion on the Peace River where the bluffs are high and confine the river in quite a narrow box. Following the Peace beyond Fort Vermilion we passed Caribou Mountain into Wood Buffalo National Park. I had always wanted to see it since my ecology professor at Harvard, Hugh Raup, had worked there and had described its ecological and biological features. It did not take us long to spot a herd of buffalo and we flew low so we could see them. They were bothered by the airplane noise and began to gallop off. It was not like the prairie grasslands where we had seen herds of buffalo being kept in reserves. These animals in Canada were really wild and they dashed off into the aspen and spruce mixed with grassland and copse. It was a treat to see the game the way our western explorers saw them in great herds. The buffalo anatomically are considered identical to our bison.

Leaving the Peace, we went on to Hay River and Great Slave Lake, a large body of water. Across the big lake is Yellowknife, one of the largest towns in this part of Northern Canada. There at the airport we were surprised to see Mary Lou Brown and Ruby Shelton, who are on the staff of Herb's research laboratory in Arizona. Mary Lou, Herb's assistant for years, is a very competent pilot, and Rub is quite a famous woman pilot. She had won the cross-country Powder-puff Derby more than once. This is the national competition for women pilots. The two ladies had flown up here in the Cessna 182 knowing we would probably be stopping at Yellowknife. I remember that well for while the four of us were together, I announced that I intended to buy an airplane. We had a nice reunion and shared a good dinner. It was quite a step for me to take, but these trips to distant places were a source of wonder.

After Yellowknife we went north over the great lake and into the far reaches of the McKenzie River. We stopped at one very small place next to the river and talked to perhaps the sole resident, learning more about the local lore and relations with the native people. We saw more of that when we landed at Inuvik, a village populated by a combination of Eskimo and northern Indians. The houses are nearly all some combination of Quonset and another form of tin-roof structure. Nearly all were connected to each other by a covered walkway that allowed one to walk from one place to another protected from the numbing cold. Perhaps it was only the common or public buildings that were so connected but the extent of these covered walks was surprisingly large.

Again I was impressed with the ubiquitous trash. Of course, in that climate nothing decomposes so even the smallest object discarded will remain there for decades. We went around the village—after all it is quite small—looking at everything. We knew nobody so we walked back to the airstrip and started the Beaver down to the end, but when we turned the plane around, the tail wheel was stuck in the snow, which was about a foot deep everywhere.

The Beaver is a heavy airplane and the tail wheel is supposed to rotate but in the snow it would not rotate, and the tail was much too heavy for the two of us to lift.

We walked the length of the runway to a large building that, as it happens, is a hangar for the Canadian Mounted Police. There was only one man there who was both pleasant and helpful. He had been a mechanic for the Mounties for some years. Somehow, with his help, we got the tail wheel free of the snow and the airplane turned around in a direction for takeoff.

Now I come to the part I do not remember, why we didn't fill the gas tanks there, but we didn't. With many thanks to the Mounted Police we took off and headed north. The Inuvik is close to the very large Mackenzie River, just about half the size of the Mississippi and half that of the Ganges, but still the 16th largest river in the world. Near its mouth it is about a mile wide and flows between surprisingly high cliffs that continue out into the Beaufort Sea for some distance. Passing over the delta we emerged onto the great expanse of partly frozen ocean and we continued out so far we could see no land, yet down on the ice there was a native with a long team of dogs pulling a sled going toward the east at a good pace.

Having a good look at this lonely expanse we retraced our steps passing by Tuktoyaktuk, which appeared to be a little larger than Inuvik but located on the coast at the north end of the many-channeled delta of the Mackenzie. After passing over Inuvik we turned westward to begin a climb over the Richardson Mountains that border the valley of the big river with what looked like a precipitous wall. For a little while we searched for a break in that wall but there seemed to be none. Turning around we now watched for a break in the clouds that hung over the cliffs. This took a little time and finally there appeared to be a slight break and we climbed into it. It seemed forever but we finally emerged above the cloud but at a considerable altitude. We did not have oxygen aboard. The compass was useless and we found that we were too far from any station to receive a signal on the radio.

The sun had been high when we started but I could see it was a little lower. Finally the clouds broke a little and we could see that the ground below us was a jumble of snow-covered peaks with dark valleys between. Time went on and I began to look at the gas gage on the auxiliary drum of gasoline and the signal was not encouraging. After what seemed a long time the mountains seemed to recede and myriad valleys appeared so we were able to drop to a lower altitude.

Herb is no alarmist but he turned to me and said, "Do you think we ought to go back?"

"Herb, I don't think there is enough gas to take us back. I think we should keep going right into the sun."

Now we descended to about 1,000 feet and saw many small rivers flowing across a broad plain.

"Can you tell which way they are flowing?" Herb asked.

After some concentrated looking, I said, "They are all iced over and the point bars do not seem diagnostic to me."

"What does the gas look like now?"

"It is not promising."

I took over the flying so Herb could devote full time on listening for a signal on the radio. There was nothing, but he kept listening to any wavelength that might be possible.

Finally I had to speak up. "Herb, I think we have about a half an hour of gas." Time passed. Herb began in a calm voice.

"I tell you what I'm going to do. At a low altitude I'm going to find a lake and stall over the water so that the tail will hit the water first."

"OK." I turned to the area behind my seat, pulled the shotgun free, then searched for the sleeping bags until I felt I had exhausted the area within my reach. We dropped down to much lower altitude and watched the flat plain ahead for a small lake.

"Herb, we have about ten minutes." In the microphone, Herb called, "Mayday. Mayday."

Suddenly he exclaimed, "I hear a signal!" Then, "I think it is from the right." We turned sharply.

"Hey, it seems stronger!" Minutes seemed to click by. Straight ahead we both saw something.

"By god, it's some radar dishes!" Still far away but getting larger, then in plain sight behind the huge dishes a couple of shacks and a strip. Herb literally dove the plane down like a stone, the end of the airstrip passed under the tail and we bounced on the ground.

At the end of the short airstrip there were some gas drums. We taxied up to them and turned off the motor. Neither of us spoke.

I got out of the plane and stood for a moment near the wheel. A man appeared out of one of the shacks, apparently an Indian, not an Eskimo. He walked up to me and said, "I glad to see you. Good luck, you be here." I nodded, rather stunned by the past few minutes and what the old man said.

"I give you good luck thing." He pressed in my hand the ivory tooth of a seal.

"You do me favor. You write me a letter. My name John. I never get letter. My daughter read it for me."

I promised, with profuse thanks. He never knew how much good luck had just fallen our way. A few weeks later I fulfilled my promise.

There was precious gasoline and we watched intently as the gallons flowed into our machine. A few hours after leaving Fort Yukon we taxied up to the signal office at Fairbanks, and were met with some sharp words. They had begun to organize a search party but had very little to go on except a sketchy idea of the direction from which the distress signal came.

Some months later I received a letter from Fort Yukon enclosing a small, beaded, circular coaster that could be used to put your wine glass on. The seal tooth has been on my leather key thong ever since.

Two Wild Rivers

I had conceived the idea of making a comparison of the knowledge gained from a river trip, a one-way traverse of a river system, with analysis of data obtained from a long record of measurement at fixed stations. I felt that there were observable facts obtained from a single pass that would be as valuable as a long period of record, measuring perhaps somewhat different parameters. By such a comparison it would be possible to evaluate the utility of a once-in-a-lifetime look at a river.

I now had a clear idea about the data needed and the mode of analysis. The same procedure would be used in a river for which there were no measurements as would be employed in a river trip on some wild but well-measured river. In both cases the river trip would be analyzed in the same manner before the published data on the well-measured data were inspected. From the previously unmeasured example as many characteristics as possible would be drawn, knowing that when compared with published data the once-over trip would provide less information than several decades of instrumental observation. But how much less would be informative. Then when nothing more was available than a single river trip, one would know in what respect the conclusions were reasonable and in what quite deficient.

The choice of a comparative and well-measured basin was carefully considered and the river chosen was the Middle Fork of the Salmon, Idaho. This flows through the Frank Church Wilderness Area, and has the alternate name, River of No Return. On the mainstem and its tributaries there are six instrumented gaging stations having a total of 147 station years of measurement to 1960. Few rivers have such a complete record of observation. One of the shortcomings of all such records in the United States is that no field observations are made of three vital parameters, bankfull stage and thus bankfull discharge, river slope, and bed material size distribution. All of these are measured in the field on our river trips. In recent years it has become increasingly useful to have a field measurement of bankfull stage and thus of bankfull discharge.

The crew I chose for this river trip included the experienced ones, Smuss, Chappie, Bill Emmett, and Bruce Lium, but I wanted to initiate some of the younger men in the Geological Survey who showed promise in science. One of the three initiates, William B. Bull, has turned out to be a leader in the field, making contributions of great value.

We gathered at the most headwater location reached by the road, Velvet Falls, where the Salmon is 150 feet wide, clear and rapid. The valley slopes of igneous rock are steep and high and generally covered with coarse, angular rubble in old talus or landslides.

Herb and I flew to the launching camp in the Beaver that carried a good part of our gear. We landed in an open place, a park-like swale covered with grass and everyone pitched in to set up camp. It was about dark when a young officer of the Forest Service came to our group and told us we had to pay a fee to be in the forest. I explained that we all worked for the United States Geological Survey and we were on official business doing scientific work. He explained that it was the rule and we had to pay. Herb spoke up and said merely, "Just send your bill to the Secretary of the Interior." I felt a little sorry for the gentleman who was just doing what he was told to do.

In the morning the two boats were launched and we swept over the riffle just below. The bottom of the river could easily be seen when the depth was 7 feet, but the average depth was 2 to 3 feet. At Dagger Falls, a short distance from our embarkation, we watched 3-foot salmon leaping the falls.

This upper part of the river is steep and bouldery so that the inflatable boats were scraped and at the end of the day Smuss was patching the tears on the bottoms. We took a complete set of measurements and then turned our attention to the steep slopes of the valley sides, making notes on the sizes and form of the common debris and talus slopes. In the evening I caught enough trout for dinner.

The next camp was on a log-strewn gravel bar backed by handsome Ponderosa pines. A chipmunk was munching on pancakes while the crew was munching on trout. There is a terrace above this camp 22 feet above the river and covered with pines, but above that steep talus slopes with sagebrush and a few spruces. The lower part of these slopes is nearly unrilled. When bedrock is prominent it gives rise to talus cones consisting of large, angular blocks that are old and black with some small, fresh, light-colored blocks near the top of the cones. The latter are obviously recent and the two contrasting types mean changes in the climate and the effect of time.

About the time we passed Marble Creek the so-called "primitive area" was getting less primitive, a great disappointment. When the final bill was passed in the Congress setting aside the Middle Fork for some protection, commercial interests and the Forest Service permitted many types of intrusions, including an airstrip, modern bridges, and tourist facilities. Despite these, it is still an interesting and more-or-less natural area that absorbs river recreation and some commercial facilities.

We camped below the airstrip near the mouth of Indian Creek not far from which there was at that time a new bridge, still in garish orange paint. Herb had flown in to give us more supplies and we had a pleasant time together. Kevin and Bill Bull climbed up to measure some slopes, Bruce and Wayne made a discharge measurement of Indian Creek, while Bill Emmett and I measured details of river character.

We were planning to camp at the mouth of Marble Creek but there was a pile of new bridge girders as well as a privy sitting in sight of the river, detracting from the feeling of the place so I decided to go on. We went past Hannah Creek and the Forest Service Guard Station with an American flag standing in a so-called primitive area. The river then became much wider and deeper with fewer rocks and falls. At this point we were just above the mouth of Sheep Creek and met up with a loud, boisterous bunch of Boy Scouts who moved in above us without as much as a with-your-leave. Sheep Creek for some unknown reason is full of green algae, the rocks being yellow in color and slimy—indeed slippery.

We measured the main river below the mouth of Tanner Falls using the boat to traverse the channel that was both wide and deep; it also had a high velocity. The hill slopes near camp were especially interesting. The talus slopes terminate at the 22-foot terrace so the terrace must be younger, and because it isolates the hillslope from the river, one wonders whether the slope continued to recede at a constant angle or changed its slope.

An airstrip nearby had a cache of supplies that Herb had flown in for us to pick up. The camp we made on the left bank was dominated by a beautiful big Ponderosa pine. The boys caught some good trout for breakfast in a big pool near camp.

At Big Creek another discharge measurement was made by boat. This tributary again had much algae, though the rock fauna seemed abundant. I found Caddis flies missing and the rocks slippery with algae. The water seemed moderately eutrophic. There were several boat parties passing us and for my money it was too crowded. We continued our measurements of hill slopes until it rained, forcing us to sit tight for a while. We then came to some of the roughest water we had seen. Porcupine Rapids put a lot of the river in both boats and in this reach one of our men fell overboard and finally reached a rock, exhausted but safe.

We had obtained a lot of field data on the river itself and, in addition, accumulated experience on hillslopes in steep terrain. In keeping with the plan, I made no attempt to analyze the data until the comparable data became available from another wild river. I was now ready to tackle the much more difficult task of obtaining comparable data in a real wilderness. Ever since Herb and I had explored by air the approaches to the big mountains of the Brooks Range, it had been my hope that we could see it on the ground, quite a different experience. It took a lot of planning because we would have to transport our boats and equipment from Arizona to Alaska. Herb and I had flown the length of the John River from its junction with the Kobuk to Anaktuvik Pass and we also knew a settler, Bill Fickus, who lived on that river with his native wife and two children. His short airstrip, carved out of the forest, would provide us an intermediate point from which to launch an expedition.

The technicians in the Phoenix Laboratory for Water Investigations had all been helpful preparing the boats and equipment, and they loaded the gear and drove it to Livengood, Alaska, in preparation for our expedition. Livengood is a small airstrip between Fairbanks and Bettles.

Herb and I flew the Cessna 182 from Phoenix to Edmonton with only a short stop at Salt Lake City. The Beaver plane was flown by Chappie, carrying Bill Emmett and Russ Brown. From Edmonton to Watson Lake, the weather was bad; we put the two airplanes in a hangar because it was both wet and cold. The following day we went on to Fairbanks. I flew the plane the whole distance that was good cross-country experience for me. At Fairbanks I got a hunting license and duck stamp and we called on the personnel of the Geological Survey.

To advance our plans, Chappie, Smuss, and I flew a reconnaissance in the 182, landed at the airstrip of Bill Fickus on the John River, and continued down the Koyukuk to Hughes. In the meantime, Herb was moving equipment in the Beaver from Livengood to Bettles. Bill Fickus very kindly offered us the use of a couple of cabins on his property, but I chose to set up a base camp at the end of the airstrip. Herb and I made the first of several trips carrying equipment from Bettles to the base camp in the Beaver.

The airstrip that Bill had carved out of the brush was about as wide as a car track, fine for his little plane, but the wings of the Beaver extended out over the brush on both sides. The Cessna used most of the length, the Beaver needed about $2/3$, but Bill's little Aeronca only about one half of the strip.

Herb and I unloaded the first of the gear near the base camp near the end of the runway and he returned to Bettles. I loaded my rifle and looked around. It was quickly apparent that a camp behind the bank of a gravel terrace would be best to get partly out of the cold wind. It was a good decision. That first night the temperature was far below freezing and the world was very wet. Even in the cold, I targeted my .270 rifle because it looked very much that I might need it.

I decided first to explore Crevice Creek to see if we could float down it to the river. It was quickly apparent that there was too little water so that alternative was not available. I wanted to go to the river but there was no trail that I could find so I bushwhacked through the brush and the hummocky marshland until I reached an expansive gravel bar beyond which was the river. On the edge of this bar I saw my first wolf track; it was like a dog's except longer, varying from 4 to 6 inches in length with nails more distinct than they would be in a dog. There seemed to be many along the river.

The John River looked uniform, wide and shallow, flowing over exceptionally well-rounded gravel of very uniform size. The stones under water were covered with a thin film of gray-green slime without any obvious evidence of organic life. I saw only two fish rise during the whole trip down the river. There seemed to be no cover for fish, no big rocks or overhanging brush, except for some deep pools, or an infrequent waterlogged snag. The tributaries have much the same character but they looked more likely to have fish because there are holes, undercut banks, and side sloughs that form a hole at the confluence. One supposed that they should support steelhead. Ducks were all gone, except for one bunch of 15 sprigs and one teal. Moose and bear must be common. When I saw a bear track and a wolf track on the same patch of mud I knew I was in really wild country.

At the campsite near the end of the airstrip I waited for Herb to return accompanied by Smuss and Bill. We built a crude, damp and cold camp behind a gravel bank. Cooking utensils did not arrive with this load and dark was closing in. We opened emergency rations from the airplane. The Arizona office had kindly prepared some canvas bed covers or bags that were waterproof. Fearing condensation I used mine as a cover for my bedroll, but Bill slept in his and found in the morning the inside was flaked with ice. A mummy bag when covered with a light canvas was good to about 17°F.

It took considerable time to transfer all the gear from Fairbanks to Bettles and then to Crevice Creek. The truck carrying equipment was driven to Livengood, some two hours north of Fairbanks, but this proved less than successful for the airstrip was both short and with high grass. The Cessna could not use it at all, and the Beaver could carry a full load only under conditions with a good headwind. As a result it took two full days to collect the gear at Crevice Creek, but we got busy making measurements, first on Crevice and then on the Allen River, near the airstrip and the camp. The third day we decided to launch at a river bar about four miles upstream of Crevice Creek, it being a difficult half mile from the airstrip to the John River. Bill Fickus flew Herb in his small plane so that he could judge whether the Beaver could land on it. He found that it was more than adequate, some quarter mile in extent, composed of well-packed fine gravel of uniform size.

The first load of equipment included two inflatable assault boats used by the Marine Corps, an outboard motor for each, gasoline, oars, and the first of the associated camping gear. It had been brought up to the camp at Crevice Creek from which Herb and I flew it to Departure Bar. The two of us put everything down on the gravel near the river and he returned to get the next load and Smuss and Bill. Herb left me on the bar with the first load of gear, for we landed on the great expanse of smooth gravel with plenty of room to spare. I sat with the equipment with my rifle on my knee and wondered how long it would take a bear to travel from the willow and spruce border to where I sat near the river.

When the whole party of four assembled on the bar, we started a measurement on the John itself using one of the boats to move across the wide channel. I surveyed a profile down the river by plane table to get the water slope. Late in the day it began to rain so Smuss floated down a half a mile and pitched a wet camp on a silty flood plain on the left bank. There was no dry wood anywhere. Chappie and I stopped our survey when it was too wet to continue, for the paper on the plane table became wet and the eyepiece of my alidade became fogged. We took the other boat and floated down to the camp. We cooked in the wet tent on a small gas stove and it was difficult to keep a fire going in front of the tent because everything was soaked.

After a miserable night, it continued to rain and we had to decide whether to move camp. I was dissatisfied with the survey of the previous day so Bill and I went back to the bar and extended the profile. Even then I was not happy with the result for we found that the river was exceptionally steep over a riffle and so flat over the long pools that the slope was close to zero.

Later in the day we took one boat down to Crevice Creek—it was too rainy for airplanes—and made a measurement on 63-mile Creek, the mouth of which is on the right bank about opposite the Allan River. Chappie and I took our rifles and walked upstream to get an estimate of bankfull stage using a hand level; we could not get a good value near the mouth because of the backwater of the John River at high stage.

Trying to take the boat upstream to camp we foundered on a shallow reach where for some distance across the channel the water was of equal depth, just too shallow for the boat to cross. I put on boots and got in the water to help the boat get across, but the boat pushed me over and I was wet above the waist and came down with a hell of a cold for it was sometime before I could get on some dry wool. In that climate it is difficult to keep both dry and warm.

The next day we broke camp and floated down to a campsite on the right bank. At six o'clock on a wet morning Chappie grabbed his shotgun and from the mouth of the tent shot a raven that fell nearly in camp against a strong north wind. I called the place Raven Bar. It was a wet camp on gravel with rocks big enough to be felt in bed through a thin mattress. This place was another lesson; one must keep his bed away from the sidewalls of a wet tent. My bed got pretty wet, at least on the outside of the roll. Suffering from a bad cold it was a damn uncomfortable place because a heavy northerly wind kept blowing all the time so we ate our breakfast fare standing up.

The following day we floated down to the mouth of Timber Creek, a particularly interesting day for we constantly were on the lookout for a bear or wolf. Chappie and I made a hunt for bear in the muskeg and spruce just east of camp where there were lots of tracks, fresh scat, and bushes loaded with berries. No luck.

We got a message from the airplane that Herb was soon leaving for Seattle, but it was clear he wanted to see us off the river before he departed. He as well as I was concerned that we might not find a gravel bar that could be used for landing and takeoff. After some days we reached the Malamute Fork of the John where we made measurements. By pressing on we reached the big river, the Koyokuk, on which there were gravel bars all right, but because of the size of the river they included coarse rocks averaging the size of footballs. Nevertheless the Beaver landed on that rough ground, but the real test was whether a heavily loaded plane could take off. There was no alternative. The boats were deflated, the gear packed, and anything unessential was set aside. As I sat in the copilot's seat next to Chappie and looked at that extensive pile of rocks, I was far from confident it would lift off, but few pilots have the experience and expertise equal to his. With more than the usual flap extension, Chappie gunned it to its maximum before releasing the brakes, and to my amazement it bumped along on top of that rough surface and lifted off. That night it was misty and cold and for some reason I now forget we decided to start for Fairbanks. One of the crew insisted on piloting the Cessna but within minutes we were in real danger and Herb finally demanded that he take over the controls and somehow got us back to Bettles. That hour was the worst in my flying experience.

Comparison of Two Wild Rivers

We had measurement data for 7 locations on the John and its tributaries, including the one measuring point on the Koyokuk.

This was to be used in comparison with the well-measured river in Idaho at the same level of detail. Though the Middle Fork was not pristine like the John, it had no dams and no appreciable forest clearing or agriculture. For each there were seven measuring points and all observations were made in the same way.

The purpose of the exercise was to determine how much information on useful parameters can be gleaned from direct field observation made only once at each spot along the river, and compare the values obtained from those computed from long-time observation on the same river. On the Middle Fork of the Salmon, we had data collected during the trip of about one week, just as we did on the John River. The river parameters most important or most useful depend on the use for which they are needed. Nowadays owing to the interest in river restoration and flood integration the bankfull stage or discharge has become prominent, but from the standpoint of water supply the average annual discharge is perhaps the most important single parameter. The variability of flow is obviously important but it still is a statistical value and is not a discrete parameter. For detailed design purposes, the values of velocity and depth at high discharge are useful.

Considering first bankfull discharge, even at measuring stations with long records it has unfortunately never been determined routinely and has to be estimated from flow frequency, and then only on an average basis. Many studies have shown that rivers reach bankfull or above with an average frequency of two years out of three, expressed as a recurrence of interval of 1.5 years. I used the value of 1.5 years to estimate bankfull discharge from the long records of stations on the Middle Fork. In contrast, bankfull stage or height and thus bankfull discharge can be clearly specified by field indicators, and techniques for making the measurement are now widely known. At this point it should be explained that, when rivers are entrenched, *bankfull* does not refer to the top of the actual lip of the trench but refers to the discharge known to be the most effective in moving sediment and reforming the channel.

With these definitions in mind, analysis of the data for the Middle Fork showed that the limited measurements in the field provide more consistent values—less scatter in the data—than the instrumental record and give values over a much larger spread of discharge quantities. On the Middle Fork the bankfull data from the stations cover a range of discharge from 2,000 to 11,000 cfs. On the other hand, the estimate of mean annual discharge for the river trip information depends on a poorly defined and limited relation of drainage area to the ratio of bankfull to annual discharge. Therefore the estimate of this parameter is very good at drainage areas greater than 2,000 square miles, whereas for basins of 100 square miles or less the error can be large. The upshot is that single-station measurements provide values for width, depth, and bankfull discharge, but are far less useful for estimating mean annual flow.

Perhaps the main conclusion is that the long instrumental records exclude some of the information now considered not only valuable but necessary. Yet the present trend is dictated by financial constraints and by failure to recognize new needs; the trend is to contract rather than expand the measurement network, a most unfortunate direction.

ID# Environmental Impact

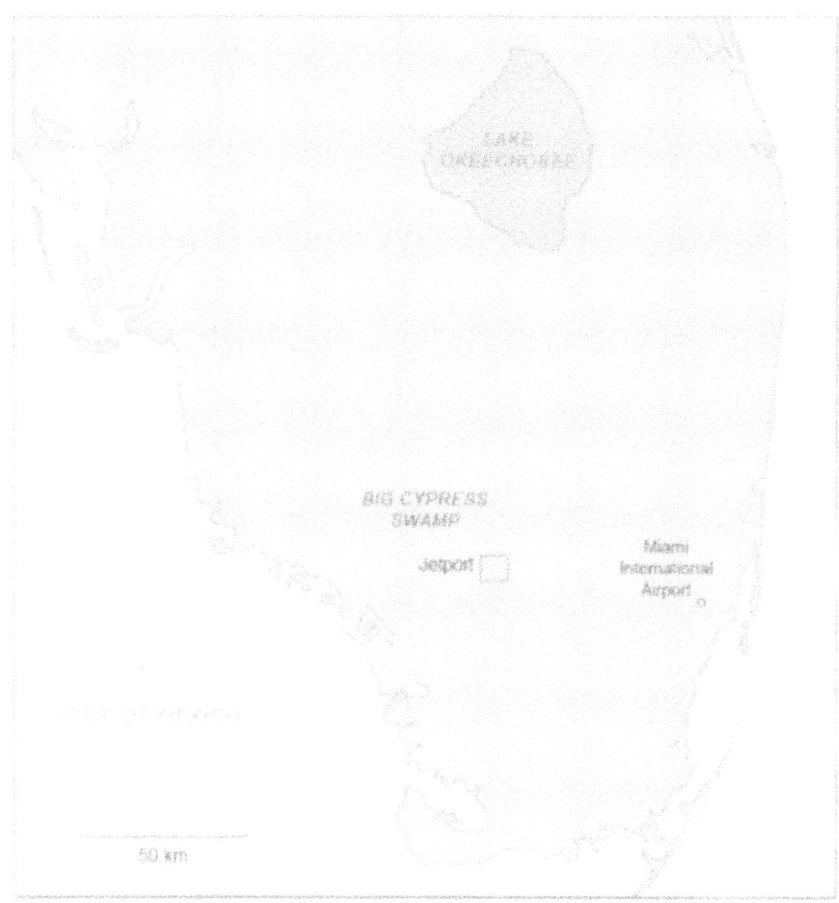

Location of proposed Florida jetport.

The Florida Jetport and the First Environmental Impact Statement

For some years when I was working in Wyoming, I did not have a house with a telephone and the local lumber yard where I bought many supplies allowed me to use theirs. One day in 1969 one of the workers of the company came out to where I was surveying and told me, "The Secretary of the Interior wants to talk to you."

A short time later I was speaking with the Undersecretary, who told me that the Department of Transportation was planning to build an airport as large as that in New York not far from Miami.

The site seemed to be somewhere close to the Everglades National Park. "I want you to go over there and investigate and prepare a joint report with Transportation on the project so I can see if it will impinge on the Park.

A couple of weeks later I flew to Florida and gathered representatives of the various offices of the Department of the Interior in the southern part of the state. I explained what was ordered and asked for cooperation and they responded in a helpful manner. Arrangements were made to fly me over the whole area, take me by airboat through the northern end of the Park, and show me the Big Cypress swamp where the airport would be built. It was a real education and I became quite familiar with the hydrology, ornithology, zoology, vegetal associations, and other aspects of the biology of the area. There were many specialists with whom I had long talks.

The problem we faced was this: In 1968, the year before the report was ordered, the Dade County Port Authority was purchasing land, mostly through condemnation, consisting of 39 square miles located in Big Cypress Swamp, 6 miles north of the boundary of Everglades National Park. The purpose was an airport for jet traffic starting with a single runway oriented east-west, to be used initially for training pilots. It was to be expanded to a full facility to relieve pressure on the large airport serving Miami. When the report was ordered, the first runway was nearing completion.

It is well known that the ecological and hydrological situation in the Everglades and the swamp areas adjoining it are unique for, in its natural state, there was a shallow flow of water covering a great swath of south Florida moving south very slowly, less than half a mile per day. This flow, called the *hydroperiod*, extended 3 or 4 months after the rainfall period. Because the land is very flat and only a few feet above sea level, the ecosystem that exists is dominated by sawgrass, which makes up 65 to 70 percent of the vegetation. Interspersed are tree islands. In the Big Cypress Swamp, large trees, open elongated forests, and large areas of stunted cypress grow in a seasonally flooded prairie.

Development of agriculture and housing north of this area has caused the water flow and its distribution to be altered by levees, control structures, and ponds, all partly contaminated with fertilizers, pesticides, and other introduced materials. So even without the airport, the hydrologic system had been highly altered and the burdens on the Park were already great.

After several fruitful days, I called the group together and we discussed the next step. Among other things I learned that the Secretary expected the report by a date less than a month away. It was also clear that everyone expected me to take the lead in the whole operation. Therefore I detailed some arrangements. Each participant was to write up a report on the subject assigned to him, the scope of each being within the area of specialization of each participant. The offices represented included the experts in fish and wildlife, birds, reptiles, mammals, hydrocycle, ecotones, parks, Indians, and others, for all of those were represented in the agencies of the Interior Department.

Everyone seemed to understand what was needed and I felt quite confident. I explained that my own work was pressing so I would come back from Wyoming in just two weeks. Then we would collate all the contributions into a report. The Department of Transportation was alerted to the plans of the Department of the Interior and when our draft was complete I would come to Washington to incorporate their contribution into a joint report before the deadline date. So without much worry I returned to Wyoming.

At the end of two weeks as stated, I returned to Florida and called the group together. When I asked to see their written contributions, to my dismay I learned that not one of them had written a single page. I was flabbergasted. I immediately ordered that they provide two secretaries who took dictation, and I needed them on a full-time basis. It was essential to have on hand someone with experience and talent in freehand drawing as well as a mapmaker. Two competent women were recruited immediately and they did their work efficiently and well. Within a day a good draftsman was available. The hour was late but I started to dictate.

The specialists had done an excellent job of teaching me so that until I found some fact that I needed from them, I could proceed without help. We were completely engrossed in dictation, typing and drafting for five days and long evenings. W. R. Marshall, Fish and Wildlife Administrator of the Bureau of Sports Fisheries and Wildlife, was of great help for he knew the details of the ecology and the geography of south Florida. Twenty-five other people repre-

The Everglades report—the first environmental impact statement.

senting eleven state and federal agencies provided technical help for the project.

At the beginning of the report, there is a two-page summary with the recommendations. There follow seven pages summarizing the environmental impacts divided into three levels of airport development from merely a training facility to full airport completion. Then the report describes in detail the physical environment and the ecosystem, including geology, hydrology, vegetation, and the natural dynamics of the ecosystem.

This is followed by a discussion of airport operation, airplane takeoff and landings, cargo handling, transportation, and land development. The threats to the ecosystem were analyzed, as were the effects of industrial growth around airports and effects on water assessed. Introduced chemicals, pesticides, and other wastes were estimated. There is a detailed discussion of bird species, their habits, movement, requirements, and their possible response to an airport. The same applies to the local species of mammals, amphibians, and reptiles, including alligators. Bird strikes are considered, as is the effect of noise on different biologic groups. The probable effect of the proposed development on the Miccosukee Tribe is discussed in detail.

In summary, the report states that the physical runway itself would not be a major problem but the infrastructure it would draw around it, including roads, buildings, water structures, pollutants, and traffic, would demolish the ecology of the National Park. The report entitled, "The Environmental Impact of the Big Cypress Swamp Jetport," is 155 pages in length, with 12 maps, 3 graphs, 4 diagrams, and 10 tables.

I carried the report to Washington and talked to members of the Transportation Department inquiring when I would receive their contributions. This went on for several days and as they provided nothing, I took their names off the report and submitted it to the Secretary on the prescribed date.

The report was circulated in government circles with not much action until the Undersecretary of the Interior, Nathaniel Reed, who was from Florida and a friend of the governor, convinced him that the conclusions were sound. When the governor repudiated the airport, the matter was settled and the plans were dropped.

The report became the model for further environmental impact statements and the requirement for such analyses became standard.

The Oil Pipeline in Alaska

When exploration drilling was allowed in Alaska near the Arctic Wildlife Refuge, there was much discussion about how the oil would be transported to the lower forty-eight states. As the prospects were examined in detail, the preferred mode would be via a pipeline extending south from Prudhoe Bay to southern Alaska, probably somewhere east of Anchorage, from where oil would be shipped by tanker to ports on the west or south coasts. It was finally decided that a pipeline would be the preferred mode and the contract finally went to a company called *Alesco*. Before work could begin Alesco was asked to determine the route for the pipeline and to provide detailed maps. The governor of Alaska at that time was Hickel, subsequently Secretary of the Interior, a man dedicated to advancing the economic position of that state.

The scale of the survey map (*next page*) was determined, the route was chosen, and Alaska was elated. Alesco recommended that the pipeline across the southern half of the Brooks Range would follow the valley of the John River, a valley that met the much larger Koyukuk not far from Bettles. Herbert Skibitzke and I knew something of this country for we had flown up that valley past Anaktuvik at the head of the pass to the Arctic Plain and had floated down the John River.

As we had seen, this or any route over the Brooks Range is in wilderness, whether so designated or not. It is truly untrammeled except for a very few families who prefer living in this distant place where there are no schools, no roads, no stores—just timber, bog, and rivers. Herb and I had got to know Bill Fickus, who built a little house on the John River and a primitive landing strip carved out of the brush. This acquaintance was a great help to us on more than one occasion.

The US Geological Survey had not been a party to all the discussions and decisions regarding the pipeline project, but when in 1969 the contractor had prepared the route maps the office of the Secretary of the Interior decided to seek the approval of the Director of the Geological Survey. The call came to me and the Office asked if I would look at the plans and maps and make a

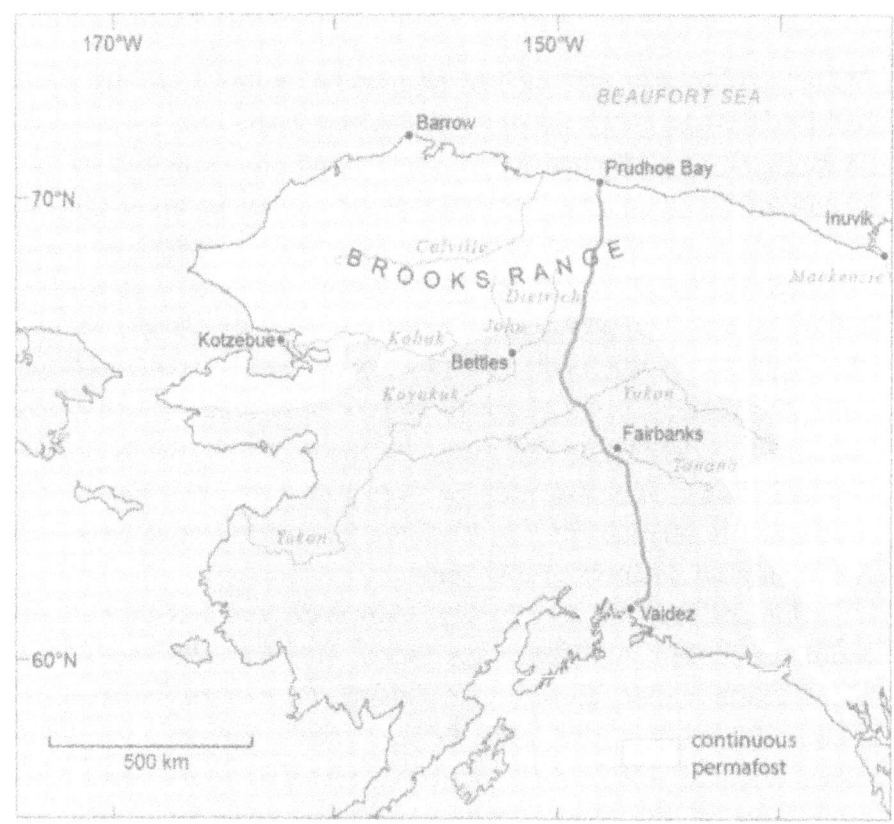

Alaska and the projected pipeline.

recommendation to the Director. I agreed to do so and a courier brought the plans to my office. The plans included folded maps as required to specify the exact route recommended by the contractor. These plans were brought into my office, not in a briefcase, not in a box, but piled five-feet high on a wheeled dolly.

I looked at this monumental pile and asked, "How much time am I given to study this?"

"Fifteen minutes," was the reply.

"Very well," I said, "sit down for the fifteen minutes and I will respond."

I took off the top of the pile the first batch that had a few pages of text as well as the first of the folded maps. I inspected the text primarily to see if they included a drawing of the cross-section of the proposed construction. There was indeed such a drawing or, at best it might be called a *sketch*. The sketch was about 3 inches wide and 2 inches in height. It depicted a round pipe, lying on a bed of gravel, in a trench with a width three times the diameter of the pipe. The pipe was covered with a thin layer of gravel. There was no scale.

This was their preparation for a multimillion dollar project? I was astounded.

"All right," I said, "take it back to the secretary's office."

I started upstairs to the office of the Director. When I could see him, I told him about the so-called report that was supposed to be the proposal for a permit to begin work. "Director," I continued, "They have apparently never heard of permafrost. The oil is warm to hot. Just imagine how the frozen soil will react. They never heard of freeze and thaw, and soil moving downhill from frost action. How about river crossings? There are lots of rivers to cross. Sir, you must reject this permit application emphatically."

"I'm sure you're right," he answered, "but I will be in a better position to tell that to the Secretary if I had some firsthand knowledge. Why don't you go there yourself, inspect the route, see what they are doing, and report back to me."

"I will get started right now. I want to use one of our own airplanes to do this and we will pilot it ourselves, and I want to choose the colleagues to go with me."

"That's fine. Get it going."

I called Herb Skibitzke, my pilot companion with whom I had traveled to Alaska, and indeed had scouted the John River where the pipeline was to be constructed. Then I called Dr. Robert Curry, a colleague who was an experienced geographer, a professor at the University of California Santa Cruz. Bob had spent years in the high mountains of the Sierra Nevada, is an expert in climatology and the effects of cold climate on the mountain landscape. His experience would exactly fit the analysis we were to make.

Herb fitted out the Beaver, the plane we had used for such expeditions, took off from Phoenix, picked up Bob and then me, and we were off to Alaska. The route we took this time was through Calgary, Dawson Creek, Fort Nelson, Teslin, and Whitehorse to Fairbanks, more or less along the Alaska Highway.

Compared with other trips we had flown to Alaska, this one was easy because the worst of winter was over and, though there was plenty of snow, the weather was decent. In Fairbanks I made some appointments to see various people both in the federal government and state officials as well as some citizens who had some interest or acquaintance with land management. We took off from Fairbanks to go first to Bettles, where fuel would be accessible, and then up the valley of the John River.

Our first foray into the field was an eye-opener. There were fresh bulldozer tracks everywhere, created by pushing over trees, brush, big rocks, or anything in the way. In these first few miles, a road was still being built, rough but apparently usable because very large trucks were using it, the kind with closed cargo holds like moving vans. In several places the incipient road had been widened and in such areas construction materials, oil drums, and other equipment were stored. The stretches of bulldozed clearing of particular interest were places the machine had crossed the river. In many instances the operator had cut his swath right up to the riverbank and, without deviating, continued to go into the water, over whatever gravel bar there was, up the other bank and into the riparian zones. But in most cases he cut his way to the riverbank, apparently looked to see if he could find a place where the bank was low, for he backed up, cut another direction, inspected the bank, and tried again. This is hardly surprising because we had experienced during our measurements at least one bank is against a terrace some 12 feet above the river, though elsewhere the banks at flood stage are some distance away from the higher terraces and only 5 feet high.

It was obvious, even from the air, that the bulldozer operator did not dismount from the machine to inspect the river bank but did his reconnaissance by trial, cutting a new swath at every try. A little knowledge of geomorphology and river channel character would have saved him both time and the riverscape.

We were seeing what we and several others called the *Hickel Highway* after the former Governor. He obviously wanted the pipeline built so the state could start the oil flowing out and the money flowing in. The road was completed, in a preliminary way, very quickly. The state of Alaska had promised an all-season road from Bettles to Prudhoe and the beginning told volumes of what would be experienced: the dissection of the Brooks Range by a barrier of unknown consequence to a variety of wildlife, especially grizzly bear;

and the rapid and irreversible dissection of swaths of lichen tundra, lying over permafrost; melting of permafrost under wheel tracks, which creates routes for rapid runoff from tire tracks to cut gullies, some of them very deep; litter, trash, oil drums, and abandoned equipment that will neither decompose nor be picked up is a perpetual eyesore. In short, it was the degradation of a fragile ecosystem unique in the Americas. In the big trucks on the new road we could see the rifles of the drivers sticking out of the windows of the cab, an indication of the future of wildlife using the highway corridor.

Because we knew Bill Fickus from previous trips we found his primitive airship in the valley of the John River near the mouth of Crevice Creek. We were greeted warmly by Bill and his Indian-Eskimo wife and we met again their two children, who had grown since our last visit. The parents are educating their two children in both normal school subjects and also in native lore, how to survive the harsh climate and the insects, and how to treat game that was an important part of their diet. We had noticed a carcass of a moose on top of a platform he had built to expose the flesh to cooling, usual for game being hung.

Over coffee, we listened to Bill talk about life in this wilderness, the several jobs he had done recently, flying a hunter to some distant place to hunt Dall sheep or black bear, landing on a gravel bar after showing his client where the animals had been seen. These excursions were but small parts of his life as wilderness guide, hunter, trapper, and fisherman, all part of feeding his family and enjoying the solitude as well as the chase.

We soon took up the purpose of our visit and his temper rose slightly. Only a few weeks ago he heard a distant motor, certainly an unusual interruption. He stepped outside just as a bulldozer emerged into his back yard, knocking down trees and scarring the fragile ground cover. The operator did not even have the courtesy to come to his door, nor to accept the offered coffee. The roaring beast proceeded up the valley without a how-dee-do. This little incident introduced the near future of the frontier. We followed the new road up the valley for many miles.

Having made copious notes and taken photos, we flew back to Bettles and then progressed to Fairbanks. The most interesting of my appointments was a discussion with the supervisor of the Bureau of Land Management for the state of Alaska. I have no transcript of the conversation but the following will give the gist and tone of the discussion. I explained that I was sent by the Director of the US Geological Survey to look at the situation on the ground and to see if the procedures being followed are in accord with the interests of the Department of Interior.

"Sir, the route of the pipeline goes through many miles of land administered by your agency, the BLM (Bureau of Land Management). Are there any restrictions or rules set up by the agency to see that values in the fields of forestry, soils, geology, and hydrology are being protected?"

"No, there is no need for any rules. We want the pipeline and whatever is needed from us we will provide."

"But there are values in wildlife, in stream stability, and in vegetation that need some attention, lest they be harmed."

"Look, we have lots of land. As far as the wildlife is concerned, we need to kill off the wolves and the grizzlies. We have land to spare. Let's use it."

This was the attitude we found in all the offices. Everyone seemed enthused by the prospect of a pipeline that would pay money to the state, the citizens, and the companies.

We returned to Arizona and I proceeded to Washington to write the report. It was a thorough document entitled, "The Environmental Impact of Oil Development, Northern Alaska." In the report, the rapidity of change was stressed. Whereas before 1968 few people even traversed the Brooks Range and there was essentially no commercial development of the north slope, that is the expanse of low-lying ground between the Brooks Range and Arctic Ocean. But within the first three months of 1969 when the proven oil field was demonstrated, the bulldozed track up the John River Valley was pushed through and truck traffic began to cross the Range. This traffic included road-building equipment, flatbed trailers, Caterpillar tractors, semis with trailers, road graders, and other machines. It was estimated that every day, when weather permitted, there were 40 flights of large aircraft transporting goods to the North Slope from Fairbanks. There were originally only two airstrips on the Slope but in 1969 there were 15 new ones.

These rapid changes occurred after little or no planning and only the most rudimentary knowledge of the conditions or processes operating there. No elementary rules, advice, or restrictions existed to help guide the development toward efficiency and responsibility.

The report discussed in some detail the characteristics of the landscape unique to that area, the permafrost, the nature and fragility of the lichen in the tundra, the subarctic forest of the taiga, the very slow rate of growth of the vegetative cover and thus the great difficulty of repairing damage inflicted by use. A great consequence of construction in this landscape is the effect of downhill slow motion of soil and rock materials due to the freeze–thaw alternation. The text included a discussion of the various values of this land, including ecological value of a stable ecosystem, recreational values, economic ones, indigenous populations, fish and wildlife, and aesthetic values. The report analyzed the projected effects of development on all these components.

Recommendations were made to the government administration that I hoped would bring attention to the particular needs and characteristics of such a high latitude, arctic land. The recommendations, as requested by the Director, went beyond just consideration of the pipeline and applied to various aspects of man's impact on an arctic environment. We made a plea for detailed study of the design of the engineering construction to ascertain how a pipeline carrying hot oil could be constructed to avoid adverse effects on permafrost, allow passage of caribou, and cross rivers without destruction of the river banks and yet maintain channel stability.

The report included a series of recommendations summarized briefly here.

1. A corridor of no shooting within two miles of the road will be declared.

2. Issue regulations governing the exploration for and construction of roads with proper enforcement, to protect the streams, the hillslopes, and native vegetation from unnecessary degradation.

3. Outlaw and enforce prohibition of shooting game and wolves from airplanes or helicopters.

4. Eliminate the bounty on wolves.

5. Require and enforce regulation of exploration, drilling, and construction to conform to standards of cleanup, debris disposal, and uncontrolled waste disposal.

6. Set standards for cutting of timber on federal land considering the fact that trees of large to medium diameter were rare.

There should follow a study of streambank stability and sediment production to learn how to minimize problems of sediment in channels. Similar study was needed in the field of engineering geology.

My report was transmitted to the Director and thence to the Secretary of the Interior with the recommendation that further construction of the pipeline be halted and that a research program be started on an improved design.

In important ways the result was gratifying. Construction of the pipeline was halted and finally deferred for about 5 years though work on the road continued. Money was provided to begin a research effort at the University of Alaska. They constructed a full-scale model of a 48-inch pipe supported on pillars that carried hot oil as it would in the field. The research results were useful for they led to a design that has proven satisfactory; the caribou were shown to be willing to go under the pipe where it was elevated on pillars. During this period of suspended construction, the route over the Brooks Range was studied in more detail, with the result that the John River route was abandoned in favor of the valley of the Dietrich River, some miles to the east. The research has paid off handsomely and the staff of the university deserves credit for providing a practical solution to a big problem.

Unfortunately, the Interior Department did not act on my subsidiary recommendations. The people in Alaska are accustomed to unlimited use of the land, without regulation, and with a plethora of wildlife that appear to them limitless. They see wolves and grizzlies as inimical and a source of income. This unconcerned attitude toward land is exemplified by the nearly universal trash to be seen beside every road, especially on the outskirts of the larger towns.

8
Environmental Ethics

Values in Conservation

Forsaking his inheritance and its assurance of a comfortable existence, Gautama Buddha adopted the life of a pauper to seek the intellectual joys of pure contemplation. Under a mulberry tree, it is said, he propounded a 12-point program of ethical conduct stressing the development of a disinterested outlook in each individual. Temples, ritual, and idols he considered distractions from the basic idea. He felt that there was a need for the development of a different point of view.

The Brahmins as well as the lower castes recognized the merits of the system suggested by Buddha, but they molded his teachings into an accessory to existing customs. They soon forgot that Gautama wanted no idols and no temples. They forgot his admonition that a simple approach was the thing that really counted. Despite his expressed wish, today Buddha in stone, in bronze, and in gold, ponders these things in thousands of temples and hears the prayers of millions who still seek the truths of an ethical life.

Today, conservation has its temples. The temples of conservation include thousands of irrigation reservoirs; it has prayer-sticks in miles of contour plow furrows, hundreds of straightened river channels, and the Buddha of a drop-inlet structure looks down on a silt-filled pool in myriad detention dams.

Environmentalism is well established today in the minds of the American public. It seems appropriate to analyze at this time just what it is that is established. In what ways have we, too, substituted the temples, the ritual, and the idols for essential values?

Most people in the American public want to have some open space, birds and most animals, parks to go to and access to such outdoor amenities. When they think about it more carefully, they recognize that conservation involves the idea of sustained yield of renewable resources and prevention of extravagant waste in nonrenewable ones. In a less specific sense it implies the preservation of values and the use of a resource for the public good through an indefinite future. In a third, and still more subjective sense, conservation connotes natural or wild things, the country and landscape beyond the confines of our own back fence.

A moment's contemplation of these concepts makes it apparent that the concepts in parts of the first definition, sustained yield and extravagant waste, are relatively objective things that can be measured and studied. Five loaves and two fishes can be weighed, tagged, and price-marked. They can be rung up on a cash register. The prevention of waste can be measured, in a way, in twelve baskets gathered full.

In the second concept, a preservation of values implies that worth must be evaluated. Use for the public good means that there should be some way of determining what is in the public interest, and what is not. This is where we begin to have trouble. There is an immediate tendency to express value in the ordinary daily-life measure, dollars. Therefore there is pressure to express one mallard as equivalent to $2.00 or to say that one goose equals $6.00. This leads to an even more spurious equation—that the value of a park is measured by the dollars spent by park visitors in the local stores. Surely a park has a far greater value than that, but expression of this greater value is not simple. In any event, the matter becomes confused when the public good is measured strictly by the number of people using a resource. Solitude and aloneness also have value.

An important element in the present-day environmental movement is the idea of the wilderness reserve. The reserve system means setting aside specific areas as wilderness and keeping these areas free from the encroachment of roads, as well as other types of development. Following the idea originally proposed by Aldo Leopold, that wilderness reserves be designated in areas of federal forest land, a system of such tracts was delineated by the Forest Service beginning with the Gila reserve in New Mexico and Arizona. Much later the wilderness reserve concept was extended to particularize different kinds of tracts, differentiating wilderness, primitive areas, and other classes of reserves.

The wilderness concept is closely allied to the concept of national parks. The park system involves two kinds of uses: so-called recreational development and, separately, areas preserved as wilderness, both within the confines of a single national park. The park system is characterized by complete protection from hunting and from the cutting or transplanting of vegetation. At the same time, the park system implies the development of roads, access trails, and, in the most accessible areas, concessions, as well as campgrounds and picnic areas. Road development and concession areas are for the express purpose of drawing in the public, and opening up such areas to recreational use. This contrasts sharply with the idea of wilderness protection, where access is limited to those traveling by foot, by pack train, or by canoe. This paradox in the administration of parks exemplifies the conflict between alternative uses of a resource, each competing use being justified by its classification as conservation.

This line of thought leads directly into the third and most subjective concept. The idea of natural, wild things is, no matter how thin you slice it, an essential element in what people think of when they speak of conservation. Sports in the woods or wilds require at least a semblance of naturalness in the setting. A natural setting contributes materially to the esthetic pleasure. Furthermore, the aspect of sportsmanship under wild conditions connotes an ethical exercise in which conscience and self-discipline play a part. When hunting and fishing are separated from these esthetic and ethical connotations, they may convey recreation but they are far removed from the concept of conservation. A baited duck blind or an illegal bag may be recreation, but gone is that restraint that is an essential element in conservation.

As one deals with those concepts that are far enough down the scale of objectivity to defy simple monetary evaluation, he is increasingly pressed to substitute visible symbols, which in themselves may not be any direct measure of conservation value, but which by a roundabout reasoning are felt to be indicative of such value. These are the idols and the prayers sticks, the temples and the trappings. If we who consider ourselves conservationists are worthy of that name, it is not too early to analyze our own attitudes critically, to ask ourselves whether the idols and the prayer sticks, to which we point with pride, have become substituted in our minds for conservation.

All human beings are easily taken in by labels, because people believe they can understand something better when it can be pigeonholed, tagged, or named. Labels make things black or white. Soon we forget what was behind the label, or how the original differentiation was made. Therefore, when one label is more popular than another, anything anyone wants to sell or promote is tagged with the popular label, and immediately becomes better or different.

Conservation is a popular label. As a result, there are plenty of programs, practices, special interests, and misrepresentations riding on the coattail of a popular movement. There is not only great attention to what something is called but also a preoccupation with visible signs—idols you might say. Things that can be seen and counted become synonymous with basic accomplishment. The number of miles of farm terracing can be counted and may be sold as a measure of how much conservation is being applied to the land. A careful scrutiny may reveal that the only land management measures that are in widespread use are those that are financially beneficial to the operator. Yet these measures may be advertised as conservation in the broad sense, and in fact they may become the only things being done on a program that is called a *conservation program*.

One of the pointed—and I might add— heartening indications that at least a few land managers have seen through the popular labels was a statement made by a Wisconsin farmer testifying before a hearing of the Water Resources Committee.

> Practices such as contour strip-cropping and terracing are very important, but I maintain that they do not get at the source of the trouble, which is soil compaction and a very low organic-matter content in our soils. It should be the goal of every farmer to restore and maintain his soils in a spongy, organically rich condition.
> I do not feel that the State or Federal agricultural agencies have given this phase of conservation work the emphasis it deserves.

This land manager indicated, in effect, that government agencies have become principally concerned with keeping soil in place. He said our concern ought to be on the quality of the soil. If we did that, the other would follow. The visible signs of so-called conservation—terraces and the like—can be measured and advertised. The basic accomplishment is more subtle. Payments or incentives, federal or state aid, may go to the man who has the nicest temples and idols to the conservation gods.

In some instances, visible signs of accomplishment of conservation, signs of having done something, may be less important than visible signs of having done nothing. When you find a piece of really fine canoe water or a little remnant of prairie flowers, you may thank your lucky stars for this visible sign that man has done nothing here.

In analyzing the field of conservation I came to the conclusion that looking at resource problems needs to allow for the existence of values that are noneconomic but nevertheless real. A conservation attitude would recognize esthetic and ethical values as distinct from and immiscible with economic ones. It would admit that maintenance of esthetic values of a resource may be incompatible with development of economic values, but would not make esthetic and ethical values compete with economic ones in monetary terms.

By such an admission, competing uses may be viewed in a framework that allows choices to be made on reasonably objective grounds rather than behind a screen of distortion and of differing definitions.

Stream pollution involves the loss of three kinds of values. First, in more extreme form, pollution may be a direct threat to public health, and health is a commodity that civilized communities do not evaluate in dollars, though for some purposes dollar valuations are placed on certain aspects. Second, pollution can be viewed in terms of the dollar cost of purification necessary for use of the water. Third, pollution as a stench and an unappetizing condition can be viewed as esthetic degradation. On the other hand, streams are polluted because economic pressures prevent municipalities and industries from adequate treatment of wastes, which must perforce be disposed of in bodies of water.

Let it be recognized at the outset that in an industrial world waste disposal is a mandatory as well as a logical use of rivers and lakes. The problem is not whether streams are to be used for transport of wastes, but what it is worth to the community to keep waste disposal within limits of esthetic and salubrious acceptability.

In a few places, the citizenry of the area decided that they wanted the stream cleaned up or freed to the sky by demolishing the concrete culvert in which it had been confined. They did not decide this because it was economical or because it would attract tourists to the local stores. Public health was not the issue. The decision was based on a desire for more pleasant surroundings in which to live, in other words on an esthetic premise. Even local industries might participate in such a venture. Local and temporary pollution by industrial waste should not be interpreted as lack of willingness to cooperate but should be accepted as a result of circumstances. In those few locations, pollution abatement may be attributed to a conservation attitude.

We Americans must decide what we want for ourselves as well as for our children. We want the advantages of a mechanized civilization, but many also want to preserve intellectual, esthetic, and ethical values. Resource development and use brings these into competition, and decisions must be made. Many of the conflicts in the use of natural resources stem from a confusion of these different kinds of values. If all resource development and use is judged by an economic yardstick alone, resources having principally esthetic and ethical values will continue to disappear as rapidly as they have in the past half century. With regard to areas of real wilderness, we are already on the flattening portion of a die-out curve that is approaching zero.

Like Buddha's followers, we seem to have forgotten that it was the viewpoint that was the important thing. We have substituted idols and the temples and have molded conservation into a trademark to help sell preconceived ideas. Let us take a new look at resource development and resource use. Development or use that is economic rather than esthetic should be compared strictly and objectively with alternative lines of action using uniform criteria for judging economic value.

For resources that are principally noneconomic in value, let us decide whether we want them, but not by assigning a dollar sign to scenery and not by making the sale of hot dogs a measure of the worth of a park. Conservation must have the positive objective of better utilization of our resources and our environment in order to make possible better and fuller lives for all the people. In this context conservation can mean development of a resource, such as water or minerals, and it can also mean preservation of a resource in its natural state such as scenic landscape or a piece of wilderness. Development or preservation would depend on which is more conducive to a better and fuller life for the people of the country.

Though such thoughts as these are foreign to the politics of the profiteer, we in the environmental community must clarify our own concepts about what are the basic principles for which we fight. To achieve a better and fuller life, we must look beyond the temple idol of monetary evaluation. We must think beyond the prayer sticks of the voluminous reports that ostensibly determine whether a given project is justifiable.

It would be less esoteric and far more honest to decide that certain elements are necessary for better and fuller lives without dreaming up ways to put meaningless dollar values on them. Let us weigh the other elements having monetary value by uniform and objective procedures, so that alternative uses of the resource can be compared and choices among the alternatives made on sound monetary grounds. We must first identify land-use and resource development measures, which are financially sound investments, quite justified without a conservation label. We would then fortify our position and consolidate our ranks when there was need to maintain or protect some esthetic value, some piece of scenery, some wood lot full of ladyslippers, some stretch of white water or of wilderness, which could never be justified on strictly economic grounds. It will require a self-restraint that can come only with the development of a conservation attitude for us to fly the conservation banner only on those things whose value lies not in our pocketbook but only in our heart.

A Reverence for Rivers

The early Greeks recognized that the direction of political and economic history of a state was materially influenced by geography, climate, and social customs. Even in the present day, the exact interaction among these cannot be accurately forecast, but the view is indisputable.

We are now in the throes of a climatic change. It surely should come as no surprise for the geologic evidence of changes in the last few thousand years is everywhere at hand. Such variation has been the outstanding attribute of Quaternary Time. Climatic change will differentially affect various geographic regions and will alter their social history in ways that are not foreseen. Though we cannot see what will transpire in the coming century, the indications point to changes of a greater scope than humans have previously known.

There are many scientists who believe a crisis is looming, though others do not see such dire results, but nearly all see changes of real importance.

I will begin with a cursory inspection of our resource base. The nonrenewable resources of greatest importance are those furnishing energy, oil, and coal. To be sure, the great deposits of coal can continue to furnish needed energy for a long time, but they also are contributing to the unsettling situation of less atmospheric shielding due to the deleterious effects of the increase of carbon dioxide.

The energy resources of oil are already on the depletion limb of the abundance curve as King Hubbert showed many years ago. Though the limitation of these are obvious to everyone, less thought has been given to the fact that mining and use of metals, phosphates, evaporites and rare minerals, though not necessarily destroyed by use, are geographically dispersed so that after use never again can be collected together in usable concentrations.

As for renewable resources, water and timber as examples, all are parts of operating natural systems that can be deranged with very troublesome results. The hydrologic system of precipitation, streamflow, sediment, dissolved salts, groundwater, and evapotranspiration is typical of a system that can be so disturbed. Moreover, such operating systems are subject to natural fluctuations resulting from climate and geography. These fluctuations can be dampened but not eliminated.

I want to look here at the general philosophy that might make our response to the coming changes relatively logical and consistent at least in the field of water resources. This does not touch on the things that obviously ought to be done immediately such as a large tax on gasoline, raising the automobile mileage, and expansion of alternate sources of energy.

In the last decade there have been in several regions of the country long-lasting drought, though some places have seen violent storms. Every time a drought occurs there is a clamor for more storage dams, tapping more extensively into groundwater resources, and more trans-basin diversions of rivers. This is accompanied by even greater competition among agriculturists, urban centers, and fish and wildlife needs for access to water. Even now this competition is intense involving legal, political, and media pressures with long delays before any settlement. Similar competition is now seen among countries and these have the prospect of even more troublesome ultimate results, probable poverty for many and war.

With regard to the United States, it is clear that practically all the water that exists is already being harnessed for some use. There is not going to be any more and probably even less if the general warming continues. The uses are, however, skewed in that there are quantities being used for purposes that are of low priority or value and some that are truly inappropriate in the light of looming shortages. It will be immediately obvious that to bring about changes from present custom is going to be difficult and perhaps impossible. Even if this is so it must begin by looking at the situation and at least see what is going to be needed.

Let us be specific. The present way water is used, is priced, and is distributed in the United States involves both inefficiencies and illogicality. The most striking inequity is the fact that irrigated agriculture uses so much more water than urban use and that 60% is consumptive, that is lost to further use. Urban withdrawals or public use constitutes 7 percent of water withdrawal, 17,000 million gallons per day compared with irrigation, 110,000 MGD. Furthermore, present methods of irrigation are very inefficient. About one third of that diverted never reaches the irrigated field.

These methods have been improved by drip irrigation devised by Israel and used in the United States for some crops such as grapes. Also spray irrigation from central wells wetting a very large circular plot is more efficient than flood or row irrigation, but the preponderance of use is by the old system.

Withdrawal for industrial use is about equal to that for irrigation but is nonconsumptive. The main result of industrial use is increase in temperature of the returned water.

In the light of these figures, the intense legal battles between agriculture and the urban, fish, and wildlife needs will no doubt continue despite the obvious inequities involved. Much of this controversy about use of water stems from the immense subsidies the federal government has persistently furnished the agricultural sector. In the western states, water provided by the federal facilities and charged to agriculture is priced at ridiculously low rates and represents only a trifle of the actual cost of providing the water. To complicate the matter even more, those farm companies that use subsidized water now consider the water to their private property, whereas the original legislation for the water development never intended that the final product was irreversibly given to private parties.

The extreme irony is the subsidy given to overabundant crops that are not vital to the country and sold overseas where products cannot compete with imports from the United States and therefore form an economic drag on developing countries.

In much of the area in western states the climate is semiarid, rainfall is low, there are long periods of less than average precipitation and rates of evaporation are high. In these areas it is fruitless to try to have merely for decoration the same vegetation as exists in the rainy east. It is only logical that as competition grows native xeric plants are going to replace green lawns typical of the Atlantic states.

There are problems of the quality of water that may assume increasing importance. Water use results in some change in water quality. Water used for irrigation is consumptive, 60 percent of water withdrawn is lost to the atmosphere. But the part returned to the ground or surface carries an increase in dissolved salts. At the lower end of the Colorado River system as water leaves the United States and enters Mexico, the saline content has been increased to the extent that its further use is limited except for irrigating salt-tolerant crops.

One reason irrigation uses so much water is that there must be applied far more than that required by the crop, because excess water is needed to carry dissolved salts down to the water table where they are beyond the reach of the plant roots. Accumulation of salts in the soil makes soil sticky when wet and become hard and cloddy when dry. Furthermore, if salts accumulate in the upper levels when plants are grown, the soil becomes too saline to grow crops and areas where it occurs are just abandoned as useless. It is well known that several ancient civilizations of large scope eventually crashed because agriculture became too withered by salinization to support the society.

There is presently a significant amount of reuse of water, probably not recognized by most people. For example, cities located in the lower end of the Mississippi system use river water that contains the effluent of the many cities upstream. This is possible because the engineering of water treatment is now so efficient that pathogens are nearly absent after treatment. The effluent from sewage disposal plants has also been greatly cleansed but treatment that diminishes pathogens does nothing to reduce the content of dissolved salts. Detergents are not taken out by treatment.

Let me summarize the types of changes that probably will be forced upon us. The allocation of existing water must become more equitable and aligned with the priority needs of the people. These will require a combination of leadership, legal adjustments, and legislative action. Price of water must more nearly reflect its cost.

Efficiency of water use must be improved, especially in agriculture considering the large proportion of the total it uses.

Water will be used in the urban sector with more concern for total withdrawal than previously, increase in native vegetation and reduction of high-demand species.

Water quality will become increasingly important as reuse of the resource expands.

In the face of the obvious limitations of resources, whether renewable or nonrenewable, continued and indefinite expansion of resource use is patently impossible. Some movement toward a steady-state condition that lies within the bounds of resource availability is not only the crux of a resource-management philosophy but is the acid test of leadership. The public is learning. It may well be the best political course to pursue.

There is a balance or harmony in natural systems, which, dictated by the laws of physics, has gradually developed during the four billion years of earth history. The maintenance of this balance is not only to the advantage of human organization, but should be the object of both our wonder and our admiration. The desire to preserve this harmony must also be incorporated into any philosophy of water management.

There is no doubt that the types of changes mentioned are going to be wrenching on the society that has been used to cheap and plentiful water. There are going to be many objections and the responses will include the following:

a) Our technology can fix it;
b) It is politically impossible;
c) Impractical idealism of crackpot environmentalists.

But crisis has the possibility of engendering an atmosphere in which new opinions as well as decisions may be molded. Crisis offers an advantage in that wide attention is attracted to the problem even if its underlying causes are perforce clouded by the immediacy of pressure. There is at least a greater probability that diverse factions might be persuaded to look more closely at their common problems and perceive what unanimity exists in their joint aspirations.

In a management philosophy and plan, it is far more necessary to minimize impact of dry years than in contending with wet ones. Though the risk of a deficient year is always present, seldom are definite plans on hand to cope with the situation when it finally arises. Rather, at the time of crisis there is a tendency toward grandiose plans to eliminate one further increment of risk, but a residual risk remains. The same crisis will occur again, less often, but equally sure. Now is the time to lay plans for muting an assured future event. It is not the time to plan expensive projects to reduce the risk by some small increment.

There are strategies that might help prepare for such eventualities. They will ameliorate the losses, but not eliminate all hardships. But, as in all water development, they require time and advance preparation. One is as follows: There are in various parts of the western states groundwater bodies too deep to be economically developed under usual economic standards, or of marginal water quality. These, and especially those remnants of the ice age—groundwater bodies presently not being recharged—should be saved from ordinary development and reserved only for the times of exceptional need. But advance engineering is needed to explore and tap them, and to connect to them transmission lines ready for some future contingency.

It is hardly in the public interest to deplete progressively stored but irreplaceable groundwater bodies whose greatest social use might be as unused reserves to be drawn on sporadically only when the need is grave. Such sporadic use would greatly extend the life of such irreplaceable water and would put it to a highly valued use.

A shortage of water will be accompanied by a call for more dams and increasing the storage capacity for those that exist. Experience in the last few decades has exposed the devastating side effects of dams, greater from large dams than from small ones. The immediate adverse effects include the displacement of people who had lived in the reservoir area, the destruction of productive land that is submerged, and a dewatering of a section of the river downstream. In developing countries the forced movement of people is usually devastating. They are nearly always moved to a place less productive and an environment less agreeable than that which they had, and it destroyed their culture, homes, and livelihood, especially if they depended on fish and water for transport.

The effects on the channel downstream of a dam are often not seen immediately but can be devastating. River channels depend on the natural fluctuations of flow for maintenance of the hydroecologic continuum and the processes of erosion, deposition, and movement of sediment are altered. Only recently have strategies been enunciated that reproduce in constricted form the most essential aspects of flow variation but very few dams are being operated to make use of these new findings. Fisheries can be decimated by dams that block navigation or by changes in time and volume of discharge.

The destructive effects are far fewer and smaller in magnitude if the dam is small and if it is located at places along the river above a tributary that will continue to contribute sediment and flow variation.

Hydroelectric development is usually an important part of the justification for dams as in irrigation, but the benefits go to a different population than that which suffers the impact.

The management of resources cannot be carried out successfully if it is looked upon as just another facet of economics, administration, and politics. Yet the latter view describes rather accurately our present approach to resource use—it can hardly be called *management*.

A philosophy of water management must pay heed to the fact that the hydrologic system is a highly interconnected plumbing network. Changes made in one part of the system have influences downstream. The continued functioning of the system is of high importance. To test whether the system is operating satisfactorily by economic and legal criteria alone will not guarantee its continued health.

In recent decades there have been in several regions of the country long-lasting drought though some places have seen violent storms. Every time a drought occurs there is a clamor for more storage dams, tapping more extensively into groundwater resources, and more trans-basin diversions of rivers. This is accompanied by even greater competition among agriculturists, urban centers, and fish and wildlife needs for access to water. Even now this competition is intense involving legal, political, and media pressures leading to long delays before any settlement. Similar competition is now seen between countries and these have the prospect of even more troublesome ultimate results, probable poverty for many and war.

With regard to the United States, it is clear that practically all the water that exists is already being harnessed for some use. There is not going to be any more and probably even less if the general warming continues. The uses are, however, skewed in that there are quantities being used for purposes that are of low priority or value and some that are truly inappropriate in the light of looming shortages. It will be immediately obvious that to bring about changes from present custom is going to be difficult and possibly impossible. Even if this is so it must begin by looking at the situation and at least see what is going to be needed.

The present way water is used, priced, and distributed in the United States involves both inefficiencies and illogicality. Commercial crops that are either surplus, of low nutritious value, and do not serve the population will be replaced by more needed varieties. At present subsidies keep these inefficiencies extant. Of course subsidies result from pressure brought on politicians. All users are going to have to pay close to the real price of water, which is a lot more than they now pay.

The management of resources cannot be carried out successfully if it is looked upon as just another facet of economics, administration, and politics. Yet the latter view describes rather accurately our present approach to resource use—it can hardly be called *management*.

A philosophy of water management must pay heed to the fact that the hydrologic system is a highly interconnected plumbing network. Changes made in one part of the system have influences downstream. The continued functioning of the system is of high importance. To test whether the system is operating satisfactorily by economic and legal criteria alone, will not guarantee its continued health. What is needed is some deeper feeling.

Speaking of the Persians, who dominated parts of Asia and most of Asia Minor in the fifth century BC, Herodotus said:

> They never defile a river with the secretions of their bodies, nor even wash their hands in one; nor will they allow others to do so, as they have a great reverence for rivers.

It is the last phrase that deserves our attention. The river is like an organism. It is internally self-adjusting, is also resilient, and can absorb changes imposed on it, but not without limit. We need something more than adjustment in laws and changes in use. Some larger view, an ethos must be adopted if the hard choices are going to be made with realism. The phrase "a reverence for rivers" might be a metaphor for what is needed if the effects of climate change are to be blunted.

The engineering capabilities of man are nearly limitless. Our economic views are too insensitive to be the only criteria for judging the health of the river organism. What is needed is a more humble view of our relation to the hydrologic system. This requires a modicum of reverence for rivers.

Let Rivers Teach Us

It is common to hear that one of the federal resource agencies plans to do fisheries habitat improvement work in a local wetlands. The improvement will consist of five log drop structures and ten log-and-rock revetments in a reach of 220 feet of channel. This, they say, will reduce erosion and the impacts of sediment deposition.

In recent decades fishermen have become more discriminating and stream managers have become more sophisticated. The angling community increasingly seeks habitat, not hatcheries, but government agencies are still building hatcheries instead of understanding. Unfortunately, many federal engineers and consulting firms have made no attempt to absorb our rapidly increasing fund of knowledge of river process and channel behavior.

Rivers do not construct drop structures. Rivers construct and maintain, by processes of erosion and deposition, channels of particular characteristics—characteristic dimensions, planforms, cross-sections, gradients, and distributions of sediment materials. These morphologic parameters are scaled to the size of the drainage basin and the nature of the rocks of the area. But they are scaled appropriately to maintain a quasi-equilibrium.

The idea of check dams or drop structures originated in the western United States in the 1930s, when the newly formed Soil Erosion Service faced the formidable gullies dissecting alluvial valleys. Experience has shown that such structures will ultimately be destroyed by undercutting, by lateral erosion of the abutments, by scour hole erosion at the toe, or by some combination of these.

If a reach of channel is suffering unusual bank erosion, downcutting of the bed, aggradation, change of channel pattern, or other evidence of disequilibrium, a realistic approach to amelioration of these problems should be based on restoring the natural combination of dimension and form characteristic of similar channels in quasi-equilibrium. These characteristics include appropriate values of width, gradient, pool and riffle sequence, length, radius, amplitude of curves and meanders, and hydraulic roughness.

A procedure might, in principle, include the following steps. Inspect the channel upstream and downstream of the reach exhibiting problems. Inspect nearby or similar valleys that appear more natural. Choose a reach of such a natural river that appears to represent the condition of the problem channel before it was disturbed or disrupted.

At this point it is useful to remind ourselves what the principal morphologic features of the river channel are that must be retained or restored. First, the slope or gradient of the channel must be the same as it is in the natural or undisturbed reach of the river. The deviation from this natural slope is the clearest reason that drop structures cannot be permanent and should be avoided.

The second imperative is the channel width. The width must represent the bankfull dimension such that when the normal bankfull discharge is exceeded, the water will overflow onto a flood plain of much greater width. This means that both width and depth at bankfull must be considered and an overflow area provided for greater discharges.

If the river curves or meanders present in the undisturbed reaches have been eliminated or importantly changed in the disturbed area, they must be reinstalled by physically constructing them. The layout of curves is the principal way the desired gradient is maintained or restored. No natural channel is straight, so the reconstruction of curves of appropriate size and shape is a main element in river restoration. The bed elevation should vary, in that pools occur in the curved reach and shallower zones in the crossover.

The dimensions of width, depth, meander, length, radius of curvature, slope, and other features have been published for many regions in the United States. These dimensions can be used as a rough check on those measured in undisturbed reaches of the river in question.

To give a few examples of such dimensions, the channel width tends to increase downstream as the square root of the bankfull discharge. The mean velocity at bankfull is, for small- to medium-size rivers, about five feet per second. A single sequence of a pool and a riffle usually has a length along the stream of five to seven channel widths. The radius of curvature for most channel bends is about two to three times the channel width. The bankfull level closely corresponds to the mean height or mean elevation of the point bar, which commonly extends into the channel from the convex bank of a channel bend.

There are a few generalizations drawn from scientific studies of channel form that can be useful in practical problems of river restoration or maintenance. Width is the morphologic parameter most easily altered by the river. If the river is deprived of some of its natural discharge, it will narrow its channel. Bank erosion usually will follow unusual or unnatural alteration in sediment supply or a change in water–sediment relation.

To change the channel slope is an invitation to continual trouble. Slope is the most conservative parameter in channel equilibrium and the original slope should be maintained. The increase in gradient is the main reason channel straightening or channelization is so destructive to river systems. Also, river curves provide an essential source of hydraulic resistance necessary for equilibrium.

We have a problem in river restoration that presently is leading to serious consequences, but is also a possible solution. The problem is lack of communication and trading of experience. As a result, successes in field restoration are little known, while mistakes are repeated indefinitely.

The Corps of Engineers has certain responsibilities in granting permits for some kinds of work on rivers. Yet various offices of that agency have totally different ideas of what works are harmful and what are beneficial. Experience does not seem to be discussed among offices, much less gathered, collated, and disseminated. If there is no central information base in a single agency, imagine the variety of practices among the several federal agencies doing river channel work: the Soil Conservation Service, Forest Service, Bureau of Reclamation, Fish and Wildlife, to name a few. In addition the state agencies do such work using engineers, fishery biologists, highway people. And there are private organizations.

There are a lot of people harming rivers. There are also people who are improving them. But we do not know who is doing what. We are all trying as best we know to do effective maintenance and improvement work, but there is no attempt to learn from each other. No doubt mistakes are repeated. No doubt success goes unnoticed.

There are many handbooks, instruction manuals, and how-to-do-it pamphlets on channel improvement. But we still lack a gradually accumulating file of case studies describing with text and illustration the original condition, an assessment of the basic cause of the problem, the techniques and construction details of treatment, and an objective analysis of the result. If such a file were initiated and all operatives urged to contribute, it is certain that we would learn from each other and our techniques would become more closely tailored to the type of river and the type of problem.

We could use an expanded effort that hopefully would involve federal and state personnel and experience. Who or what organization should take the lead is not specified. But one thing seems clear. We must let the river teach us. Not just a few of us. Let the river teach all of us.

Ethos and the Earth's Resources

It was 431 BC, the end of the first year of the Peloponnesian War, a conflict that lasted 27 years. The Athenian general and respected leader Pericles spoke to the assembled citizens concerning their strength, weakness, and prospects against the better prepared military-minded Lacedaemonians. He pointed out that the inherent advantage of democratic Athens over the rigidity of Sparta lay in the character of its citizens, who were accustomed to freedom in their government and in their ordinary lives. Pericles described this inherent character as follows: "(The) ease in our private relations does not make us lawless as citizens." The chief safeguard, he noted, is that citizens obey the customs and the laws "whether they are actually on the statute book, or belong to that code which, though unwritten, yet cannot be broken without acknowledged disgrace."

Pericles was a member of the government, in a post filled by an annual election. He was also a general, but in this speech he was speaking for the government. He described, if you will, an ethos, or guiding belief, of both government and citizen. That ethos defined the character of Athens—character of both the government and the governed.

Unfortunately, democratic government is fragile. The freedom it permits can be used by demagogues and thus can be subtly but quickly subverted into oligarchy. In Athens, it was particularly precarious. The citizens disavowed Pericles's view of a defensive war and accepted Cleon's urging for a war of conquest. Athens no longer represented the "hopeful and creative force whose humanity, though imperfect, far surpasses the repressiveness of Sparta." Athens itself became the repressor. In arrogance and conceit, Athenians attempted to conquer Syracuse and subdue all of Sicily. They failed. The ethos of Athens, its character and tone, had been eroded and finally obliterated.

The unwritten assumptions of fairness, equity, and the common good gave way to domination by a narrow class and special interests. Athens lost its strength and hope at Syracuse, lost the war despite its overwhelming advantages, and deteriorated to a second-class power.

To use the words of Barbara Tuchman, this ancient history is a distant mirror. It is useful to look in that mirror.

It is obvious that the United States does not exhibit the ethos that one visualized in the word. Perhaps a useful approach to a discussion of our status is to look at several segments of the society to see where and under what circumstances an ethos of life exists. I will start with the individual and his family.

Nearly all Americans have a desire to do the right thing, to be good citizens, and good neighbors. These sentiments are laudable but are only a part of a life ethos, a character or tone that involves fairness, equity, and consideration of others even under conditions of stress. There are innumerable examples of this attitude at the individual or family level. Many families take someone less fortunate under their wings and give financial and moral support as a matter of compassion and desire to help a neighbor. The same generous family may not see its needed role in the larger society because it may be misled and influenced by the media, by television, by politicians and demagogues, leading it to lend its vote to poorly understood promises or being frightened by scenarios drawn by special interests.

There is an area of family life, sports hunting, in which an ethos still exists though it is gradually being codified. There is a spirit in the hunting community that gives precedence to skill rather than trophy, or restraint rather than the kill. This attitude began in the late 1800s when hunting as a sport, not a necessity for food, became common. There were excesses to be seen, as in the extermination of the passenger pigeon. My grandfather, who lived in Burlington, Iowa, was an avid hunter who, with his eldest son, Aldo, took an early-morning train across the Mississippi River to hunt quail in the breaks of the great river. He could see even in 1890 that it would be better for the game and for the hunter if there were no shooting during the nesting period of spring. So without the requirement of law he advanced the idea in his family and his colleagues. Later this concept was entered into laws and regulations, but the modern sportsman recognizes the value of this spirit of restraint. Part of the hunting ethos involved skills in outdoor living, and a gradual appreciation of aesthetic values in scenery and in wild creatures within their habitats.

There were other forces that led to the extirpation of the grizzly, the wolf, and the passenger pigeon. Driven by money there were excesses in timber cutting and destruction of forests, but Theodore Roosevelt sensed these from personal experience and started forest reserves and eventually a national park system.

These examples are somewhat distant from the concept of ethos because many restraints are no longer personal and freely chosen but are codified and in regulations and law. But the source or desire did precede or encouraged the restraint by regulation.

Hobbies, such as gardening, bird watching, whale watching, and now butterfly watching, are perhaps the best modern examples of mass recreation that stems from an aesthetic that is uplifting and nondestructive.

At one time the spirit of the game underpinned most team sports though the rapid emergence of competition with large monetary stakes, excessive salaries, advertising, and large rewards for winning have deprived sport of one of its most valuable assets. In the major sports now, the spread of dishonesty and corrupt practices have become worse with drugs and steroids.

Government and the Loss of Equity

The ultimate loss is experienced when government itself loses a sense of equity, fairness, and honesty. The founders of the republic expressed their ethos in the brilliant words Jefferson wrote in the Declaration of Independence, but the pragmatists, such as John Adams, were sensitive to the probability that the fragility of any ethos would overcome their brave words. So by long bargaining they agreed on the checks and balances of a tripartite form of government as a basic tenet of the Constitution. And just as they had foreseen, history shows how the attitude of government has surged and ebbed in its acceptance of the concepts of equity, fairness, generosity, and honesty. At the turn of this century we are caught in the lowest segment of that cycle.

My concern here is that the resources of the earth are vulnerable to lack of restraint in the exercise of power. It is a well-established fact that the natural world, its processes, its productivity, and diversity provide humans with the essential ingredients of life, but it is equally established that when these are degraded or undermined they do not recover either quickly or completely. Added to that changes in conditions on earth constantly impinge on these processes and conditions, changes that cannot be accurately forecast and may become semi-permanent, at least in the time span of civilizations. Among these are climatic change, sea level, composition of the atmosphere, quality or composition of radiation from the sun, and earth movements.

Because all forms of life and most processes affecting the habitable earth are dependent on or are influenced by water, its form, presence, quantity, and quality are influenced by all of the factors listed previously and through it, humans are affected as well. Therefore it is pertinent to examine some of these interrelationships and evaluate the probable direction of how they will fare over the time span of a few generations.

In nearly every part of the world the water situation is increasingly in trouble. There is a general degradation in quality, availability is limited or even quite absent for masses of people, and competition for water is already critical especially where water source and water need are in different countries. Even in the most advanced and wealthy countries, competition is a major problem.

Water needs in many regions put several of the following against one another: agriculture, urban use, hydropower, fish and wildlife, commercial fisheries, and recreation. A major complication is the construction of dams to serve some of these purposes at the expense of others. Even countries lucky enough to be plentifully supplied with water are now racked with these competitive forces, the United States being an example. But in less wealthy countries, especially in Africa, water availability is a major factor in life as demonstrated by statistics. One hundred percent of the population in the United States has access to water for both urban use and for sanitation. Worldwide, nearly one billion rural people lack drinking water and two billion lack water for sanitation. Most of these people are in Asia and Africa.

But we in the United States have our problems, too, if of a different sort. Examples abound of individuals who have a personal interest in the water needed for their own operations, livelihood, and home use, but perceive no larger responsibility except to be prudent in water use. If citizens think about water at all, they feel they can rely on those persons they elect or on an organization they support to take the broader view. Among these are the federal and state agencies and individuals who have no direct monetary interest in the water they administer, but whose attitudes, policies, and clientele determine in great part where, how, and how much of the water resource is utilized.

The proliferation of public agencies dealing with water has led to a dissociation of their policies, their procedures, and their outlook from the operational health of the hydrologic system. Everything one entity does affects many other entities, yet each entity operates as if it alone were the flower facing the sun. There is no guiding belief, no ethos, involving the natural world. There is no concern for the common—as Garrett Hardin expressed it—no overriding responsibility for the whole.

The individuals and agencies that manage water resources are, after all, the product of the public that supports them. In Athens, it was the character of the citizenry that was expressed in the ethos of both the government and the governed. Citizens can become so divorced if they are not informed, if they do not get honest information, if government itself desires to seek its own ends. When the governed do not see the consequences of neglect of the general welfare, and if they are given no insight into the operational details of how their own interests are being handled, they are manipulated rather than informed. What both government and governed now lack is the ethos or gut feeling that the resources of the planet and the nation are worthy of husbandry—indeed, are essential to our long-term well-being.

In America there is a shocking lack of equity, of dedication to fairness, of a desire to consider various interests and treat all with some measure of equality. We see all around us the pressure of special interests, and the bending under that pressure. Equity was part of the Athenian code. As Thucydides expressed it, "Praise is due to all who . . . respect justice more than their position compels them to do."

Agencies and individuals who manage resources have functions that are described by law, but not necessarily specified by law. Thus, there is a wide range of administrative discretion not only permitted, but allocated by legislative bodies. Indeed, this is as it should be, but often lacking is a guiding precept that public service means service to the whole with a sense of balance and equity. These agencies and individuals often are under pressure from self-seeking forces. As a result, dedicated public servants are captured by the history of the organization in which they work, and are subjected to conflicting demands. The result is that ethos and equity are not part of the system.

The immediate answer to this observation swells to a clamor in my ears—the agencies are doing just what the legislature has ordered them to do. But anyone who has been in government service knows that the administrative discretion is as wide as a barn door. This very discretion permits the possibility of buckling under the pressure of outside interests. Ethos and equity are needed to promote the public interest and resource use.

We need a guiding direction for water agencies and the individuals who work in those agencies. In phrasing this guiding direction, a distinction should be made between ethos and policy: Policy can be written in explicit terms and can be in the form of an order, but ethos is less explicit and includes viewpoint—a guiding value that is understood but not necessarily written out.

The Hydrologic Continuum

An essential component of such a guiding direction should include the following idea: decisions in the field of water development and management should aim toward the preservation of the integrity of the hydrologic continuum: The hydrologic cycle continues for good or ill, but the idea of a continuum implies a maintenance of balance—an operational quasi-equilibrium among the processes within the hydrologic cycle. The hydrologic continuum might alternatively be called the *hydrocycle*, involving air, water, soul, biosphere, and people.

By hydrologic continuum, I mean the effective operation of these forces in the drainage basin that maintain a balance among the processes of rock weathering, soil formation, water and sediment delivery to stream channels, and the exit of water and sediment from the basin. These forces are both biological and physical. Each part of the system modifies other parts. The flood plain reduces flood peaks. River curves maintain hydraulic resistance and help moderate velocity. Thus, both form and process interact to insure that no part of the system accelerates beyond the limits of flexibility. This is what is meant by *quasi-equilibrium*.

The integrity of the hydrologic continuum is implied in the term *sustainability*. But, in choosing to use the word, I mean more than the ability to exist through time. I mean the dynamic flexibility to adjust constantly through changing circumstances. The integrity of the hydrologic continuum must include adjustment to a changing climate by gradual noncatastrophic alterations. But the new parametric is the anthropogenic effect on the planet's atmosphere itself.

Up to the recent century, the natural resilience of the hydrologic continuum has absorbed the imposed changes and recovered, but that flexibility now appears to be strained too far as indicated by the recent increased rate of glacial melting, the record values of measured temperature, violent storms and long droughts, the extinction of species, and loss of diversity. Whereas in previous times the overuse or alteration of natural processes affected a locality, a state, or even a country, now the troubles are on a world scale.

I list a few of these related to water. The oceans are being depleted of the most valuable and numerous species; rivers of the world are impacted by more than 47,000 large dams—of which China, with 22,000, and the United States with 6,500—have the most. Tropical rain forests are being decimated; irrigated area has dramatically increased in recent years. Between 1965 and 1999 in developing countries alone the irrigated area rose from 40 to 67 millions of hectares or 1.8 times, and in developed countries from 109 to 207 million hectares, or 1.9 times, or nearly double.

As mentioned earlier, competition for water engenders disputes between countries, many of them precipitating war. Peter Gleick has compiled a list of such conflicts in the world, the number of which increased by about 60 between 960 and 2001. As he notes, "Conflicts may stem from the drive to possess or control another nation's water resources, thus making water systems and resources a political or military goal. . . . Conflicts may arise when water systems are used as instruments of war, either as targets or tools."

Looking at the world as a whole, we see an unappreciated and unintentional decimation of the factors maintaining the functions of natural ecosystems. Despite the obvious troubles of resource use in poor or developing countries, the United States, the richest and most advanced, exacerbates the already serious situation by a willful, intentional and premeditated assault on natural resources to fulfill present desires of business and moneyed interests, to sustain only for a few years an advantage over others who also must use the planet's bounty.

America should be the leader, rather than at the moment, the spoiler. What we need is an ethos that once was the hallmark for at least a portion of this society, the desire to be the first in caring for those less fortunate, for being willing to work with others to advance the good of all, to be the leader in careful use of our resources, and willingness to be one among many, part of the biologic community, not above it or disdainful of it. We must examine our relation to the earth and acknowledge the need for a land ethic with a view to altering our present trajectory.

The recent trends in loss of resources worldwide exacerbated by the climatic change now being experienced, point to a growing inability of the earth ecosystems to sustain the burden of humanity. If this support continues to weaken, the United States will be the biggest loser. Our country is rigging its ships for the voyage to Syracuse.

Bibliography

Sources are listed here for readers who wish to look at the published data.

Bryan, Kirk. 1925. "The Papago Country." *U.S. Geological Survey Water Supply Paper* 494.

Ehrlich, Paul R., and Donald Kennedy. 2005. "Millennium Assessment of Human Behavior." *Science* 309: 562–63.

Fox, Robin Lane. 1980. *The Search for Alexander.* Boston: Little Brown.

Herodotus. 2003. *The Histories.* Aubrey de Selincourt, Trans. New York: Penguin.

Gleick, Peter H. 2003. *The World's Water 2002–2003.* Washington, DC: Island Press.

Leopold, L. B. 1969. "The Rapids and the Pools, Grand Canyon." *U.S. Geological Survey Professional Paper* 669: 131–45.

———. 1969. *Environmental Impact of Oil Development Northern Alaska.* Washington, DC: U.S. Dept. of Interior.

———. 1969. "Landscape Aesthetics." *Natural History* 78(8): 37–45.

———. 1992. "Base Level Rise: Gradient of Deposition." *Israel Journal of Earth Sciences* 41: 57–64.

———. 1994. *A View of the River.* Cambridge, MA: Harvard University Press.

———, and John P. Miller. 1956. "Ephemeral Streams." *U.S. Geological Survey Professional Paper* 282A.

———, and Walter B. Langbein. 1966. "River Meanders. *Scientific American* 214: 60–70.

———, and Herbert E. Skibitzke. 1967. "Observations on Unmeasured Rivers." *Geografiska Annaler* 49(A): 247–55.

———, and William B. Bull. 1979. "Base Level, Aggradation and Grade." *Proceedings of the American Philosophical Society* 123 (3): 168–202.

———, and Claudio Vita-Finzi. 1998. "Valley Changes in the Mediterranean and America and Their Effect on Humans." *Proceedings of the American Philosophical Society* 142 (1): 1– 17.

Lovins, Amory. 1977. *Soft Energy Paths.* New York: Harper Colophon Books.

Mackin, J. Hoover. 1948. "Concept of the Graded River." *Geological Society of America Bulletin* 59: 462–512.

McBain, S. M., and William J. Trush. 1997. *Trinity River Maintenance Flow Study, Final Report.* Hoopa, CA: Hoopa Valley Tribe.

Platt, John R. 1964. "Strong Inference." *Science* 146 (3642): 347–53.

Powell, John Wesley. 1875. *Exploration of the Colorado River of the West.* Washington, DC: Smithsonian Institute.

Scranton, R., and Joseph Shaw. 1997. "Changes in Relative Sea Level." In *Kenchreai, Topography and Architecture: Eastern Port of Corinth,* edited by R. Scranton, J. Shaw, and L. Ibrahim. Leiden, Netherlands: Brill.

Stearns, H., and G. A. McDonald. 1942. "Geology and Resources of the Island of Maui, Hawaii." *Hawaii (Terr.) Division of Hydrography Bulletin* 7: 1–222.

Thucydides. 1951. *The Peloponnesian War.* New York: The Modern Library.

Trush, Willian J., Scott McBain, and Luna B. Leopold. 2000. "Attributes of an Alluvial River and Their Relation to Water Policy and Management." *Proceedings of the National Academy of Sciences* 97 (22): 11858–63.

Vaughn, F. E. 1970. *Andrew C. Lawson: Teacher, Philosopher.* Glendale CA: Arthur H. Clark Co.

Von Schelling, Hermann. 1964. *Most Frequent Random Walks.* Schenectady, NY: General Electric Report 64GL92.

Index

A
aesthetic (values), 3, 4, 55, 113, 184, 217, 239, 240
AID (Agency for International Development), 84
Alaska, 72, 181–93, 197, 211–18
Alexander the Great, 143, 147
　and the Gedrosian desert (Makran), 152–53
Allen, Smuss, 128–30, 171, 176–81, 194, 199, 200
Anderson, Clinton, 82
Amazon River, 120
Aral Sea, 142
arroyo, 9, 11, 59, 100, 102, 104, 106, 110, 149, 158
Athabasca River, 181
Athens, 238, 242, 245
Auchter, E., 22, 23, 58, 111, 112

B
Bacon, Francis, 113
Badlands, 120, 125
base level, 9, 157–59, 161–63
Batisse, M., 135
Beer, Charles, 20
Bjerknes, Jacob, 20, 21
braided (river), 15, 71, 77, 84, 94
Brodie, Hugh, 34
Brown, Carl, 13
Brown, Russ, 197
Bull, William, 160, 194, 196, 199
Bureau of Reclamation, 21, 124
Burnett, Bill, 176
Bryan, Kirk, 11, 58, 59, 61, 114, 117, 149

C
Carter, Walter, 32, 33
Central Water and Power Commission (India), 86
Chalcidice (Xerxes's) canal, 151
Chapman, Howard, 171, 178, 179, 181, 194, 197, 200, 201
China, 50, 97, 99, 244
Clean Water Act, 84
climate change, 123, 161, 234
Colorado Canyon, 125–32, 157, 170–80
Conservation Foundation, 83
Corps of Engineers, 237
cotton, 142
Cyrus, 150

D
dam, 2–3, 83, 124, 148, 157–59, 162, 163, 227, 232–35, 241, 244
Darius, 150
Darwin, Charles, 113, 114
Davies, G., 44
Davis, William Morris, 158
Delhi, 86
Denny, Charles, 82

E
Emmett, Bill, 117, 120, 125, 126, 128, 130, 131, 171, 174, 194, 197
entropy, 169
"Environmental Impact of the Big Cypress Swamp Jetport," 209–10
environmental impact statement, 206–10

ephemeral streams, 11, 15, 85, 99, 100–02, 161
erosion, 11, 21, 22, 34, 35, 46, 61, 79, 91, 99, 100, 102, 104, 106, 107, 128, 141, 149, 150, 157, 158, 162, 174, 177, 232, 234–36
ethos, 2, 234, 237–40, 242, 243, 245
Evenari, Michael, 133

F
Fickus, Bill, 197
flood control, 14
The Flood Control Controversy, 83
Florida Jetport, 206
flume, 90, 93, 164, 166
freshwater lens, 47

G
Ganges River, 84, 85
Gates of Lodore, The, 119, 121
geomorphology, xv, 4, 79, 102, 104, 105, 110, 115, 124, 145, 156, 157, 161, 163, 164, 170, 180, 214, 170, 180
Ghyben–Herzberg lens, 47, 49
Gilbert, G. K., 163, 164
glacier, 61, 67, 68, 72, 167, 181
Green River, 70–79, 119–25, 129
groundwater, 1, 46
Gulf Stream, 167
gullying, 9, 45, 46, 59, 72, 123, 128, 141–43, 161, 162, 235

H
Hack, John, 82
Haible, Will, 160
Haleakala, 44, 52
Happ, Stafford, 12
Hatch, Buzz, 120, 125, 128
Hawai'i, 20–55
  Pineapple Institute 23, 24
Herodotus, 150, 234
Holocene, 61, 149
Horner, W. W., 12, 13

Horton, R. E., 12, 13
Hunt, Charles B., 125, 126, 129, 159
hydraulic geometry, 76, 79–83, 110, 111, 159

I
India, 83, 84–98, 134, 141, 144, 145, 147, 152
Indus River, 85, 152
Inglis, C. C., 119
irrigation, 50, 52, 53, 71, 90, 95, 141, 142, 147, 149, 220, 228–29, 232, 244
Israel, 162, 228

J
Jefferson, Thomas, 111, 112, 240
John River, 184, 185, 197, 198, 199, 200, 201, 202

K
Karakum Desert, 142

L
Lamb, Harold, 137, 144
Langbein, Walter B., 21–22, 125, 130, 131, 169, 170
Lawson, Andrew C., 114
Leopold, Aldo, 9, 10, 22, 49, 66, 74, 83, 112, 221, 239
Leopold, Barbara, 160
Leopold, Carl, 112
Leopold, Estella, 171, 178, 179
Leopold, Nina, 112
leper colony, 36
Lewis, Meriwether, 111
Lium, Bruce, 171, 179, 180, 188, 194
Lyons, 23, 33, 45

M
McDonald, G. A., 54
Mackenzie R., 191
Mackin, J. Hoover, 163, 164
Maddock, Thomas, Jr., 11–14, 82, 84

Matthews, G. H., 119
Meade, Bob, 120
meander, 64, 71, 77, 79, 94, 77, 79, 101, 159, 160, 164–70, 174, 178, 182, 183, 185, 235, 236
Mesa Verde, 161, 163, 164
Middle Fork (Salmon River), 194–95, 202–03
Miller, John P., 59, 60–70, 71–75, 110, 158
Mississippi River, 83, 84, 88, 181, 182, 191, 230, 239
Mojave Desert, 16
Molokai, 34, 35, 44
Myrick, Bob, 128, 131, 167, 171

N
Nabateans, 148–49, 162
Nace, Raymond, 133, 134, 135, 136, 145, 146
National Academy of Sciences, 58
National Parks, 4, 189, 207, 221, 225, 239
Navajo, 9, 134, 183, 187
Navajo Experiment Station of the Soil Conservation Service, 8
Navajo Reservation, 158
Negev, 148, 162–63
Neiburger, Morris, 20
New Mexico, 59, 73, 82, 117
Nightingale, Gordon, 25, 26, 40, 41, 43
Nile River, 132
Nolan, Thomas, 157

O
Osterkamp, 117
Otsuka, George, 38, 39

P
Pakistan, 85, 152
Pericles, 237–38
Peterson, H. V., 22
Pineapple Research Institute, 24, 51

plane table, 8, 9, 77, 78, 103, 171, 173, 176, 200
Platt, John R., 115
Playfair, John, 163
Popper, Karl, 114
Powder River, 61, 110
Powell, John Wesley, 119–24, 157, 158, 171

R
Raup, Hugh, 189
Reiche, Parry, 10
research strategy, 44, 79, 110–11, 114, 115, 116
River Field Book, The, 188
river restoration, 202, 236, 237
Rogers, David, 177
Rossby, Carl Gustaf, 58
Rubey, William, 76
Rubin, Meyer, 171, 179
runoff, 12, 115, 123, 159, 162, 215
Russia, 132, 133–36, 144–46, 181

S
Samarkand, 142–44
San Juan River, 74, 159, 160, 171, 172
A Sand County Almanac, 22, 28, 49
Schwartz, Charles, 22, 28
Sherman, L. K., 12
Shidel, Clarence, 16
Skibitzke, Herb, 125, 127, 130, 131, 171, 177, 180–84, 192, 194, 196–99, 201
SOB rapid, 124
Stearns, H., 54
Sternberg, Hilgard O'Reilly, 120
Syracuse, 238, 245

T
Tartars, 143
Tashkent, 135–45
terrace
    agricultural, 149–50

terrace *(continued)*
   alluvial, 59, 61, 63, 73, 76, 23, 120, 122, 124, 128, 147, 149, 161, 162, 175, 178, 195, 196, 198, 214
Thomas, Harold, 137
Timur, 143–45

U
Ulugh Beg, 144, 145
UNESCO (United Nations Educational, Scientific and Cultural Organization), 132–33, 135, 136
   Committee on Arid Zone Research, 133
   Hydrologic Decade, 133
unit hydrograph, 12
United States Geological Survey (USGS), 21, 52, 58, 81, 132, 133, 171, 176, 194, 195
   Water Resources Division of, 58, 82, 156
uplift, crustal, 151, 160
Uzbekistan, 135–45

V
Vigil Network, 117, 157

W
Watts Branch, 71, 80, 117
Webb, Jean, 49
wells, 146–48
Wentworth, Chester, 47, 48, 54
wilderness bill, 82
Wind River Range, 2, 75, 167
Wolman, Reds (M. G.), 71, 72–75, 173
Wyoming, 2, 59, 61, 62, 76, 80, 110, 166, 167, 188, 206

X
Xenophon, 151
Xerxes, 150, 151

Y
Yeo, Herbert, 8, 9

Z
Zahnheiser, Howard, 82, 83

www.ingramcontent.com/pod-product-compliance
Lightning Source LLC
Chambersburg PA
CBHW050740110426
42814CB00006B/312